MATERIA MEDICA AND THERAPEUTICS

An Introduction to the Rational Treatment of Disease

J. MITCHELL BRUCE,

M.A. ABERD., M.D. LOND.,

Fellow of the Royal College of Physicians;
Physician and Lecturer on Materia Medica and
Therapeutics, Charing Cross Hospital;
Assistant Physician to the
Hospital for Consumption, Brompton.

MAVEN BOOKS

Chennai   Trichy   Tirunelveli   New Delhi

MAVEN BOOKS

An Imprint of **MJP Publishers**

ISBN 978-93-87867-66-6 **MAVEN Books**

All rights reserved No. 44, Nallathambi Street,
Printed and bound in India Triplicane, Chennai 600 005

MJP 754 © Publishers, 2019

Publisher : **C. Janarthanan**

PUBLISHER'S NOTE

The legacy of a country is in its varied cultural heritage, historical literature, developments in the field of economy and science. The top nations in the world are competing in the field of science, economy and literature. This vast legacy has to be conserved and documented so that it can be bestowed to the future generation. The knowledge of this legacy is slowly getting perished in the present generation due to lack of documentation.

Keeping this in mind, the concern with retrospective acquiring of rare books has been accented recently by the burgeoning reprint industry. MAVEN Books is gratified to retrieve the rare collections with a view to bring back those books that were landmarks in their time.

In this effort, a series of rare books would be republished under the banner, "MAVEN Books". The books in the reprint series have been carefully selected for their contemporary usefulness as well as their historical importance within the intellectual. We reconstruct the book with slight enhancements made for better presentation, without affecting the contents of the original edition.

Most of the works selected for republishing covers a huge range of subjects, from history to anthropology. We believe this reprint edition will be a service to the numerous researchers and practitioners active in this fascinating field. We allow readers to experience the wonder of peering into a scholarly work of the highest order and seminal significance.

MAVEN Books

To

RICHARD QUAIN, M.D., F.R.S.,

CHAIRMAN OF THE PHARMACOPŒIA COMMITTEE OF THE GENERAL
MEDICAL COUNCIL,

ETC., ETC.

THIS WORK IS DEDICATED,

IN ADMIRATION OF A LIFE SPENT IN THE INTERESTS OF

MEDICINE AND THE MEDICAL PROFESSION,

AND IN GRATEFUL ACKNOWLEDGMENT OF CONSTANT

PERSONAL KINDNESS

DURING A VALUED FRIENDSHIP OF

MANY YEARS.

PREFACE.

THIS book is chiefly therapeutical in its scope, and is intended to be a rational guide to the student and practitioner of medicine in the treatment of disease. At the same time the MATERIA MEDICA has not been sacrificed. On the contrary, it will be found to be set forth in detail by the adoption of a *natural* and concise arrangement, which presents the subject in such a form that it can be quickly appreciated and easily remembered. The author attaches importance to the plan which he has adopted in the description of the Special Therapeutics, and which consists in systematically tracing the physiological action and uses of the different drugs in their passage through the body, from their first contact with it locally until they are eliminated in the secretion. In the part of the manual devoted to General Therapeutics he has further departed from the ordinary arrangement, by discussing the actions and uses of remedies, not under the headings of artificial groups, but of the physiological systems of the body—digestion, respiration, etc., so as to conduct the student from facts with which he is familiar to the great principles of treatment. In using the book, the first year's student is recommended to confine his attention to the Materia Medica proper; and under the action and uses of the drugs, to read only the words printed in thick type.

The author gratefully acknowledges the valuable assistance which he has received in the preparation of ·the work from his friends Dr. Quain, Dr. Lauder Brunton, and Dr. Frederick Roberts; from his brother, Dr. William Bruce of Dingwall; from Mr. Woodhouse Braine, who kindly sketched the section on the use of anæsthetics; and especially from his friend and former class-assistant, Mr. A. C. N. Goldney, who has relieved him of much labour by superintending the pharmaceutical portions, drawing up lists, and compiling the index.

The many standard treatises on Materia Medica and Therapeutics in this and other countries have been freely consulted, especially Nothnagel and Rossbach's "Arzneimittellehre," Husemann's "Arzneimittellehre," the works of Wood and Bartholow, and the useful volumes of Squire and Martindale.

CONTENTS.

MATERIA MEDICA AND THERAPEUTICS.

INTRODUCTION.

MATERIA MEDICA AND THERAPEUTICS relate to the use of drugs in the treatment of disease. The place which these subjects occupy in the Medical Sciences lies, therefore, between Chemistry, Botany, Anatomy, and Physiology on the one hand, and Medicine and Surgery on the other hand; whilst they stand side by side with Pathology, the other stepping-stone from the more purely scientific to the more strictly practical portions of professional education. The student will now be able to turn to account his acquaintance with chemistry and biology, and to appreciate the fact that these sciences are the true foundations of all professional knowledge; and when he has reached the end of the volume he may anticipate with some confidence a personal introduction to the treatment of disease.

Let us consider what subjects are comprised under the title, " Materia Medica and Therapeutics."

Materia medica.—This term is applied to the materials or substances used in medicine, their names, sources, physical characters, and chemical properties, the preparations made from them, and the doses in which they may be given.

Therapeutics relates to the treatment of disease, the word signifying healing, from $\theta\epsilon\rho\alpha\pi\epsilon\acute{u}\omega$, *I attend, heal,* or *treat.* It includes, therefore, all that relates

B—8

to the science and art of healing, not merely by the application of the materia medica to the treatment of disease, but by the use of remedial measures of every kind, including diet, climate, baths, clothing, nursing, and the numerous other means which may be combined to restore health, not the least important being surgical treatment, or surgical therapeutics. This definition is manifestly far too comprehensive for our present purpose, which is concerned only with medicinal therapeutics, *i.e.* the uses of the materia medica. When this subject is discussed under the head of each article of the materia medica, as it comes before us in natural order, it is known by the name of the **special therapeutics** of that article. Materia medica and special therapeutics will constitute the first part of the work.

When the numerous and complex facts of special therapeutics are collected and examined, certain great principles may be educed from them, unfortunately still very far from being perfect, but sufficient to furnish the ground-work for a science of **general therapeutics.** This portion of our subject will be considered in the concluding part of the work.

Certain other terms, variously related to the preceding, must here be defined :

Pharmacodynamics (φάρμακον, *a drug*, and δύναμις, *power*) is a convenient name for that part of our subject which relates to the action of drugs upon the healthy individual, that is, the **physiological action of drugs.** In the first part of this work the term "action" will simply be used to express the same meaning.

Pharmacology (φάρμακον, *a drug*, that is, either a medicine or a poison) is a term which has been employed in two senses. With the older writers in this country it is the science that relates to the chemical and physiological properties of drugs, their selection and preparation, the extraction of their active principles, and the combination of these with others. More recently pharmacology has come to be used in a wider sense, and to include the whole subject of materia medica and therapeutics, for which it is a short and convenient term.

Pharmacy is the name applied to the *art* which corresponds with the *science* of pharmacology, the art of making the preparations indicated or ordered by the pharmacologist, and of dispensing the combinations prescribed by the therapeutist. In such a work as the present, the details of pharmacy must be mainly omitted. They have to be learned practically in the dispensary or pharmaceutical laboratory, not by rote from a book.

The Pharmacopœia.—The number of drugs used from time immemorial is enormous, and comparatively few are now believed to be really useful. To separate the valuable *materiæ medicæ* from those supposed to be worthless, books have been published from time to time by the governments or medical authorities of different countries, which furnish an authoritative list of the drugs generally recognised and used by the profession, and the preparations made from them, which have thus become "officinal" or official. These books are known as pharmacopœias ($\phi\acute{a}\rho\mu\alpha\kappa\text{o}\nu$, *a drug*, and $\pi\text{o}\iota\acute{\epsilon}\omega$, *I make*). In this country we have the British Pharmacopœia, which provides us with a tolerably accurate list of the drugs and preparations in use at the time of its publication. But as pharmacology is a rapidly-advancing science, especially from the direction of chemistry and pharmacodynamics, and as opinion is very unsettled on the subject of therapeutics, the pharmacopœias of different countries differ greatly ; and the pharmacopœia of any given country neither is accepted at the time of its publication as perfect in itself and to be followed as an article of

faith, nor remains a correct representation of professional opinion for any great length of time. It is,
however, an invaluable medium of communication
between the physician and the pharmaceutical chemist,
whom it furnishes with formulæ for a great variety
of preparations of definite composition, and an im-
mense amount of information respecting drugs which
is necessary in combining these, or in devising fresh
preparations.

Plan of the Materia Medica.—In the Pharmacopœia the materiæ medicæ and their preparations
are arranged alphabetically for convenience of reference, but in a systematic treatise they must be discussed in natural order.

The following plan will be adopted in these
pages :

PART I.—THE INORGANIC MATERIA MEDICA.

Group 1. Alkalies and Alkaline Earths.
    „   2. Metals.
    „   3. Metalloids.
    „   4. Acids.
    „   5. Water.
    „   6. Carbohydrates and other Carbon Compounds.

PART II.—THE ORGANIC MATERIA MEDICA.

Group 1. The Vegetable Kingdom
    „   2. The Animal Kingdom.

Each article will be discussed under several distinct
and definite headings, which are as follows : The
names of the drug, in Latin and in English, its *chemical
formula*, if any, and the *definition of its nature ;* its
source ; its *characters ;* its *composition ;* its *doses ;*
and the *preparations* made from it.

A general reference must here be made to each of
these headings.

Names, nature, and sources of drugs.—These
are sufficiently indicated by the above plan in the case

of the inorganic materia medica. It includes many of the chemical elements, and a great variety of compounds of the same.

Vegetable drugs are derived from entire plants, including fungi and lichens, stems (woods), green tops and twigs, roots and rhizomes, barks and leaves, buds, flowers, parts of flowers and flowering tops, fruits and seeds ; and various vegetable products, including fixed and volatile oils, resins, oleo-resins, balsams, gums, gum-resins, inspissated juices and secretions. The animal materia medica includes entire animals, portions of animals, and products yielded either during life or after death.

The methods for obtaining the drugs will generally be given, and must be learned by the student, who should repeat for himself as many as possible of the easier processes. Most of these are already familiar to him in chemistry, such as *solution, filtration, evaporation, crystallisation, precipitation, decantation, sublimation, distillation, destructive distillation, digestion,* and *washing.* A few specially pharmaceutical processes will, however, require to be defined :

Pulverisation, the powdering of drugs, is done on a large scale in powerful drug-mills. On a small scale it may be done by simple *trituration (triturare,* to pound), or powdering in the dry state ; by *levigation (levigare,* to make smooth or fine), or rubbing down with the aid of a little fluid, the resulting paste being afterwards dried ; or by *mediate pulverisation,* in which some very hard substance or medium is mixed with the drug, in order to break up its substance thoroughly. Powdered drugs necessarily require *sifting.*

Elutriation (elutriare, from *eluere,* to wash out) consists in diffusing an insoluble powder in water, allowing only the heavier part to settle, and decanting the fluid ; allowing this again to settle for a longer time, so as to deposit a second or finer size of powder, and again decanting; and repeating the operation indefinitely until an extreme degree of fineness has been reached.

Lixiviation (*lix*, a lye) is a process of washing an ash or crude mixture of solids, for the purpose of dissolving out the constituents in the form of a lye, or water impregnated with salts.

Maceration and *Percolation* are described under *Tincturæ* (page 15).

Characters.—This part of the description must be studied practically. Using the Manual as his guide, the student must examine specimens of drugs, and note respecting each article its *general appearance* to the eye, whether liquid, solid, crystalline, etc.; its *colour*, its *weight*, its *smell*, and its *taste* (if non-poisonous). If convenient, his examination of the drug should follow the pharmacopœial account farther, and include the determination of its *reaction;* of its *solubility* in water, alcohol, ether, oils, etc.; and of the *effects of heat* on its volatility, fusibility, etc. Other important chemical properties, bearing on its pharmaceutical applications, may have to be studied, especially its *incompatibility* with other drugs, which prevents their combination in preparations. Along with the characters, in many instances, certain *tests* are given, which introduce the student to the subject of

Impurities, and the methods of distinguishing substances so like each other as to be very readily confounded. Impurities may be the result of the imperfect selection, preservation, or preparation of drugs, including chemical decomposition of every kind; or of fraudulent adulteration. Similarity is, of course, a matter of accident, but may give rise to serious error.

The tests of purity applied to *inorganic* drugs are mainly such as are familiar to the student of chemistry; and to avoid constant repetition, the most common of them will be represented here once for all :

	Impurity.	*Detected by.*
	Water.	Bibulous paper; dampness; loss of weight by heat.
	Organic matter.	Colour.
	Sulphuric acid.	White precipitate with $BaCl_2$.
1. Impurities derived from the sources of the drug, or formed in the process of manufacture and imperfectly removed.	Hydrochloric acid.	White precipitate with $AgNO_3$.
	Phosphoric acid.	Yellow precipitate with $AgNO_3$, soluble in HNO_3 and in NH_4HO.
	Carbonic acid.	Precipitate with lime-water; effervesces with acids.
	Sulphurous acid.	Zinc and HCl yield H_2S.
	Nitric acid.	H_2SO_4 and $FeSO_4$ give a brown ring between the two fluids.
	Lime.	White precipitate with oxalate of ammonia or with CO_2.
	Arsenic.	Yellow precipitate with H_2S.
2. Impurities derived from the apparatus used.	Metals, especially lead, iron, and copper.	Precipitates with $(NH_4)_2S$, or H_2S; and special tests.
3. Insufficient strength.		Volumetric test
4. Fraudulent adulterations.	Various colored earths.	Non-volatility; insolubility in HNO_3.
	Cheap salts.	Various tests.
	Starch.	Blue colour with iodine.
	Sugar.	Evaporation; quantitative test.

In the case of *organic* drugs, impurities are chiefly to be detected by careful physical examination and special quantitative tests.

Composition.—The composition of the inorganic drugs is expressed by their name and formula. On the other hand, the organic drugs are frequently highly complex, the chief proximate principles being the following : Fixed oils, volatile oils, resins, oleo-resins, gums, gum-resins, balsams, pectin, alkaloids, acids, neutral substances, glucosides, starch, sugar, cellulose, albuminous substances, ferments, colouring matter, salts, and extractives.   Some of these demand general consideration.

Fixed oils are extracted by expression (if possible, without the aid of heat) from the seeds or fruits of plants, or from animal tissues.   They are composed of oleate, with palmitate and stearate of glyceryl ; that is, are compounds of fatty acids (oleic, palmitic, and stearic, as well as of other, less common) with the radical glyceryl, C_3H_5.   With caustic alkalies or metallic oxides, they form soaps, the metal displacing the glyceryl, which is hydrated, and becomes glycerine, C_3H_53HO.

Volatile Oils ; Resins ; Oleo-resins ; Balsams.— Volatile oils are obtained by distillation from entire plants, flowers, fruits, or seeds.   Most of them are colourless when pure, and highly aromatic.   They consist of a liquid hydrocarbon or *elæopten*, generally isomeric or identical with terpene, the hydrocarbon of oil of turpentine, $C_{10}H_{16}$; and of an oxydised hydro-carbon, usually a solid body, or *stearopten*, like camphor, $C_{10}H_{16}O$.   A few volatile oils contain sulphur and nitrogen.   Further oxydation converts a portion of volatile oils into *resins*, solid, brittle, non-volatile bodies, and thus gives rise to *oleo-resins*, which can be broken up into their two constituents by distillation.   Resins or oleo-resins yielding benzoic or cinnamic acids are called *balsams*.

Gums are exudations from the stems of plants. They consist of two rather complex carbohydrates, *arabin*, $C_{12}H_{22}O_{11}$, and *bassorin*, $C_{12}H_{20}O_{10}$, which play the part of acid radicals, and exist in gums as salts of magnesium and potassium. Arabin is soluble in water; bassorin is not soluble, but swells into a gelatinoid mass. *Pectin*, vegetable jelly, $C_{32}H_{40}O_{28}, 4H_2O$, occurs in a few medicinal plants, and, like the *mucilage* yielded by several others, is allied to gum. *Gum-resins* are natural or artificial exudations from plants, containing various proportions of gums and resins, or more frequently of gums, resins, and volatile oils.

Alkaloids are - active principles formed within plants, which resemble alkalies in turning red litmus-paper blue, and form salts with acids. As a rule, they are crystalline solids, rarely liquids; sparingly soluble in water, but readily in alcohol, the solution being intensely bitter.

Organic acids of great variety exist in plants, combined with the inorganic bases, such as potash and lime, with alkaloids, or possibly free.

Neutral substances are a very large and mixed group, including the carbohydrates, such as starch, sugars, gums, etc. ; albuminous bodies, which occasionally act as ferments ; a few bitter principles ; and *glucosides*.

Glucosides are chiefly neutral bodies, capable of being decomposed in the presence of water into glucose and a second substance, different in each instance.

The remaining constituents of organic drugs do not call for special notice.

Dose.—The Pharmacopœia suggests the limits within which the different substances and their preparations may be safely given to an adult. These must be carefully learned. The principles of dosage will be presently discussed.

Preparations.—The list of preparations made from the drug, with the principal ingredients, strength, and doses of each, will conclude the account of its pharmacy. This subject demands special consideration here.

Most of the materiæ medicæ possess such characters that it is absolutely necessary to prepare them for administration. Thus, if we take, as examples, Sulphur, one of the elements; Potassii Iodidum, a crystalline salt; Chloroformum, a liquid compound of chlorine and formyl; Colocynthidis Pulpa, the dried pulp of a fruit; Jalapa, a tuber; and Cantharis, a dried beetle; it is manifest that few of these can be brought into useful contact with the body in their native form. Preparations must be made from them, and for several reasons we must have *a variety of preparations*. First, as we have just seen, substances are very various; secondly, a substance may contain several active principles, soluble in different media, which it may or may not be desirable to extract together or separately; thirdly, we constantly wish to obtain *combinations* of drugs, so as to increase, diminish, or otherwise modify the action of each, or to obtain combined action; fourthly, we must provide for variety of administration or application, externally or internally, to act on a part or to enter the blood by any of the methods of exhibition to be presently described; and we must be ready to meet the tastes and fancies of patients with respect to pills, powders, etc., as well as the necessities of circumstances.

The following is a list of the different kinds of preparations in the British Pharmacopœia. A complete list of each will be found in the synoptical tables at the end of the volume.

Aceta, Vinegars, are extractive solutions in acetic acid (not vinegar).

Aquæ, Waters, are very weak simple solutions of volatile oils in distilled water, obtained by distilling the vegetable products or the volatile oil. Aqua

Camphoræ is a solution without distillation. Aqua Chloroformi is the only aqua not made from an oil.

Cataplasmata, Poultices, are familiar external applications. They generally contain linseed meal as their basis.

Chartæ, Papers, consist of cartridge paper coated with an active compound much like a plaster.

Confectiones, Confections, conserves, or electuaries, are soft pasty-looking preparations, in which drugs, generally dry, are incorporated with syrup, sugar, or honey.

Decocta, Decoctions, are made by boiling vegetable substances in water from five to twenty minutes. All decoctions are simple, except that of aloes and one of the decoctions of sarsa.

Emplastra, Plasters, are external applications which adhere when applied to the body, and produce either a local or a general effect. The basis in all is a compound of fatty substances (resin, wax, lead, soap, etc.), and is intended to be spread on linen, leather, or other material.

Enemata, Enemas, injections, clysters, are liquid preparations for injection *per rectum.* The basis is generally mucilage of starch or water.

Essentiæ, Essences, are solutions of volatile oils in four parts of rectified spirit, *i.e.* are ten times the strength of the ordinary spirits.

Extracta, Extracts, are preparations obtained by evaporating either the expressed juice of fresh plants, or the soluble parts of dried drugs. They are, therefore, of several kinds

1. *Green extracts.*—The juice pressed from the bruised plant is heated to 130°, to coagulate the green colouring matter, which is strained off and reserved. The fluid is next heated to 200°, to coagulate the albumen, which is separated by filtration and rejected. The filtrate is now evaporated at 140° to a syrup, the green

colouring matter returned, and the whole evaporated down to the required consistence.   Ex. : Extractum Aconiti.

2. *Fresh extracts* are prepared like green extracts, but there being no colouring matter, the juice is heated at once to 212° Fahr. to coagulate the albumen, filtered, and evaporated at 160°.   Ex.: Extractum Taraxaci.

3. *Aqueous extracts* are prepared from drugs by the action of cold, hot, or boiling water on dry drugs, and subsequent evaporation to a proper consistence. Ex. : Extractum Calumbæ, Extractum Gentianæ.

4. *Alcoholic extracts* are prepared by the action of rectified spirit, rectified spirit and water, or proof spirit on dry drugs, and evaporation to a proper consistence.   Ex. : Extractum Physostigmatis.

5. *Ethereal extracts* are prepared in various ways ; viz. (*a*) By percolating with ether and evaporating the product : Extractum Filicis Liquidum.   (*b*) By making an alcoholic extract, macerating this in ether, and evaporating : Extractum Mezerei Æthereum.   (*c*) By washing the drug free from oil, by percolation with ether, before making an aqueous or alcoholic extract : Extractum Ergotæ Liquidum.

6. *Acetic extract.*—The only extract of this kind, Extractum Colchici Aceticum, is made like a fresh extract, but acetic acid is added to the crushed corms before expression, and evaporation is arrested whilst the mass is soft.

7. *Liquid extracts* are prepared by macerating the drug in water, evaporating to form a concentrated solution, and adding a little spirit to prevent decomposition. Ex. : Extractum Pareiræ Liquidum.   The process is modified in the case of ergot and filix mas, as described under ethereal extracts.

The consistence of extracts varies much.   Some are liquid ; four are solid, viz. those of aloes (2), hæmatoxylum, and krameria ; five are soft, viz.

the acetic extract of colchicum, and the extracts of cannabis indica, mezereon, nux vomica, and physostigma; the rest are of the consistence suitable for forming pills.

Glycerina, Glycerines, are solutions of substances in glycerine. They are suitable either for further solution or for application locally. o

Infusa, Infusions, are obtained by steeping vegetable substances in water, generally near the boiling point. The infusions of calumba and quassia are made with cold water; those of chiretta and cusparia with water at 120° Fahr. Those of orange and gentian are compound; that of roses contains acid.

Injectio Hypodermica, Hypodermic Injection, is a strong aqueous solution of an active drug for administration with a syringe and needle under the skin.

Linimenta, Liniments, or embrocations, are preparations suitable for application by rubbing, anointing, or painting. All liniments contain either camphor, oil, or soap.

Liquores, Solutions proper, consist of substances other than volatile oils dissolved in water; but the preparations of many are complicated, solution being assisted by spirit, acids, ether, lime, other salts, or carbonic acid as in the effervescing liquores.

Lotiones, Lotions, or washes, are solutions or mixtures for external use by washing or on lint. The British Pharmacopœia contains but two lotions, Lotio Hydrargyri Flava, and Lotio Hydrargyri Nigra.

Mellita, Honeys, are fluid preparations containing a large proportion of honey.

Misturæ, Mixtures, are made by rubbing up various substances in water, the product being not a solution, but a mixture only. The insoluble substances are generally suspended in the water by means of gum, spirit, or milk. They are frequently compound.

Mucilagines, Mucilages, are solutions of colloid substances in water.

Oleum, an Oil, is a solution in a fixed oil. Ex.: Oleum Phosphoratum.

Pilulæ, Pills, are soft easily divisible masses, variously composed of extracts or substances naturally tenacious, with suitable "excipients," such as treacle, confection of roses, or powdered liquorice. They are almost all complex. The substances best adapted for giving in pill form are such as are not conveniently given in fluid form, or those intended to act slowly.

Pulveres, Powders, are compounds of dry substances reduced to powder and intimately mixed.

Spiritus, Spirits, are either simple or complex. *Simple spirits* are solutions of colourless substances or oils in rectified spirit, the latter of the strength of 1 in 50. Ex.: Spiritus Chloroformi, Spiritus Cajuputi. *Complex spirits* are prepared in a special manner; *e.g.* Spiritus Ætheris Nitrosi.

Succi, Juices, are the expressed juices of the fresh plants, to which one-third of their volume of spirit has been added to preserve them. Limonis Succus, Rhamni Succus, and Mori Succus, are not preparations, but natural products.

Suppositoria, Suppositories, are solid conical bodies, composed of active ingredients and various mixtures of fats and wax, or starch and soap, adapted for introduction into the rectum, where they are intended to melt.

Syrupi, Syrups, are fluid preparations containing a large amount of sugar.

Tincturæ, Tinctures, are solutions in spirit, either alone or combined with other solvents. They may be grouped according to (1) the *solvent,* (2) the *process,* or (3) the *ingredients.*

1. *Solvents.*—Rectified spirit is chiefly used when the substances contain resin or volatile oil, as in cannabis

indica. Proof spirit is adapted when the substances are partly soluble in water, partly in spirit, as in most tinctures. Ammonia is employed in the ammoniated tinctures of opium, valerian, quinine, and guaiacum; spirit of ether in Tinctura Lobeliæ Ætherea; and tincture of orange in Tinctura Quiniæ.

2. *Processes.*—(*a*) *Simple solution* or mixture. Ex.: Tinctura Ferri Perchloridi. (*b*) *Maceration.* Macerate the drug in the spirit for seven days; press, if necessary; strain; and add sufficient spirit to make one pint. Ex.: Tinctura Opii. (*c*) *Percolation.* Pour the spirit on the drug packed in a percolator, and add spirit slowly until one pint is collected. Ex.: Tinctura Zingiberis Fortior. (*d*) *Maceration and percolation.* Macerate the drug for forty-eight hours in part of the spirit; then percolate, adding more spirit as required; press, filter the products, mix the liquids, and add spirit to one pint. Ex.: Tinctura Digitalis.

3. *Ingredients.*—Tinctures are either simple, or compound, *i.e.* contain more than one active substance. Ex: Tinctura Benzoini Composita.

Trochisci, Lozenges, are dried tablets of sugar, gum, mucilage, water, and one or more active ingredients, uniformly divided or previously dissolved.

Unguenta, Ointments, are mixtures of active substances with lard, benzoated lard, suet, wax, or oil, variously combined; or with simple ointment. The ingredients are either thoroughly mixed or melted together.

Vapores, Inhalations, are preparations administered in the form of vapour or gas, disengaged on the union of the ingredients.

Vina, Wines, are solutions of drugs either in sherry (ex.: Vinum Ipecacuanhæ), or in orange wine (ex.: Vinum Quiniæ).

The following kinds of preparations are in common use, but are not ordered in the British Pharmacopœia:

Collyria, Eye-washes.

Gargarismata, Gargles, liquid preparations for application to the fauces.

Linctus, Linctuses, thin confections to be slowly swallowed in small doses to affect the throat.

Pessi, Pessaries, a small variety of suppositories for administration *per vaginam.*

Weights and Measures: Signs and Symbols.

The **weights** of the British Pharmacopœia are the grain, *granum;* the ounce, *uncia;* and the pound, *librum;* with their conventional symbols, gr., ℥, and lb., respectively.

The **apothecaries' scale** runs thus:

1 grain = *granum,* gr. i. ;
437·5 grains = 1 ounce = *uncia,* ℥j. ;
16 ounces = 1 pound = *librum,* lb.i.

It is very common, however, although not officinal, to employ a weight between the grain and the ounce, for the sake of convenience, called the *drachm,* ʒ, to signify 60 grains; not, let it be observed, the $\frac{1}{8}$th part of an ounce, as in the fluid measures.

A 20-grain weight, called the *scruple,* ℈, was formerly in general use, but is now mostly discarded.

Measures.—The measures of the British Pharmacopœia and their symbols are the minim, *minimum,* min., or ♏; the fluid drachm, *drachma fluida,* fl.dr., or *f*ʒ ; the fluid ounce, *uncia fluida,* fl.oz., or *f*℥ ; the pint, *octarium,* O ; and the gallon, *congius,* C.

The scale is

1 minim   = min.j., ♏ j.
60 minims   = 1 fluid drachm, fl.dr.j., *f*ʒj.
 8 fluid drachms = 1 fluid ounce, fl.oz.j., *f*℥j.
20 fluid ounces = 1 pint, O j.
 8 pints   = 1 gallon, C j.

Relations of Weights to Measures.—

1 minim is the measure of 0·91 grain of water.
1 fluid drachm ,, ,, 54·68 ,, ,,
1 fluid ounce ,, ,, 1 ounce, or 437·5 grains of water.
1 pint ,, ,, 1·25 lbs., or 8750·0 ,, ,,
1 gallon ,, ,, 10 lbs., or 70000·0 ,, ,,

Metrical system.—The metrical or decimal system of weights and measures, which is officinal on the continent of Europe, may possibly come to be adopted in this country, as being in many respects preferable to the other :

Weights —
1 milligramme = the thousandth part of 1 gramme = 0·001 grm.
1 centigramme = the hundredth ,, ,, = 0·01 ,,
1 decigramme = the tenth ,, ,, = 0·1 ,,
1 gramme = weight of 1 cubic centimetre of water at 4°C.
1 decagramme = ten grammes = 10·0 grm.
1 hectogramme = one hundred grammes = 100·0 ,,
1 kilogramme = one thousand ,, = 1000·0 ,,

Capacity —
1 millilitre = 1 cub. centim. = the measure of 1 grm. of water.
1 centilitre = 10 ,, = ,, 10 ,,
1 decilitre = 100 ,, = .. 100 ,,
1 litre = 1000 ,, = ,, 1000 ,, (1 kilo.)

Relation of the weights of the British Pharma-copœia to the metrical weights.—

1 pound = 453·5925 grammes.
1 ounce — 28·3495 ,,
1 grain — 0·0648 ,,

and conversely :

1 milligramme — 0·015432 grain.
1 centigramme — 0·15432 ,,
1 decigramme = 1·5432 ,,
1 gramme — 15·432 ,,
1 kilogramme = 2 lbs. 3 oz., 119·8 gr. = 15432·348 gr.

Relation of the measures of the two systems to each other.—

1 gallon = 4·543487 litres.
1 pint = 0·567936 ,, = 567·936 c. centim.
1 fluid ounce = 0·028396 ,, = 28·396 ,,

1 fluid drachm = 0·003549 litre —   3·549 c. centim.
1 minim   — 0·000059 „   —   0·059   „

and conversely ·

1 cubic centimetre   — 15·432 grain measures.
1 litre = 1 pint 15 oz. 2 drs. 11 min. = 15432·348   „

Domestic measures.—A teaspoonful is a convenient but not quite accurate measure of 1 fluid drachm; a dessert-spoonful, of 2 fluid drachms; a table-spoonful, of half a fluid ounce; a wineglassful, of 1½ to 2 fluid ounces; a teacupful, of 5 fluid ounces; a breakfastcupful, of 8 fluid ounces; a tumblerful, of 10 to 12 fluid ounces. Wherever accuracy is desired, a graduated measure glass must be used. Some " drops " being twice as large as others, it is specially dangerous to order drops of powerful remedies for children.

Action and uses of drugs.—The preceding subjects complete the information furnished by the Pharmacopœia; but the student must next make himself acquainted with the action and uses of each drug, that is, its pharmacodynamical and therapeutical relations. In the following pages this portion of the subject will be discussed under four distinct heads, according to the order in which the drug affects the different parts of the body. These are as follows :

1. **Immediate local action.**—When a medicine is applied to an exposed surface, it may produce some effect or " act upon " it. This may occur either *externally, i.e.* on the skin or exposed mucous surfaces, such as the conjunctiva, anterior nares, vagina, etc. ; or *internally*—on the alimentary canal, especially the stomach and intestines, including the rectum. Some drugs have no further action.

2. **Action in or on the blood.**—The great majority of active remedies are absorbed into the blood, and enter into the composition of its plasma, much less

frequently of the red or white corpuscles; that is, have an effect *in* it, but little or no effect *on* it. The student must carefully note the fact, that very few medicines produce their characteristic effect by acting upon the blood.

3. **Specific action.**—Leaving the circulation, drugs enter the tissues and organs, alter the anatomical and physiological state of one or more of them, and are then said to have a *specific* action upon these. In most instances this is the characteristic and most important part of the action of the drug.

4. **Remote local action.**—Medicinal substances, having passed through the tissues, are finally cast out of the body by the excreting organs, whether in the same form as they were admitted, or as the products of decomposition in the system. The kidneys are the great channel of escape for drugs; the lungs ("breath"), skin, bowels, mouth, mammary gland, and all mucous surfaces and wounds, to a less extent. Whilst thus passing through the excreting organs, the active principles of drugs frequently exert a further or remote local effect upon them, not infrequently resembling their immediate local influence.

Prescribing.—When the practitioner desires to employ drugs for the purposes of treatment, he turns to his knowledge of the action and uses of the materia medica, selects his remedies, and proceeds to order one or more of them, according to a recognised form or formula, which is called a *prescription.* This is a very difficult proceeding when first attempted, being nothing less than a serious and probably sudden practical test of one's acquaintance with an enormous subject. The beginner should know, therefore, what points are specially to be kept before him under these circumstances. Briefly, they may be said to be the following:

1. **Selection of the remedy.**—This is, of course, the first and fundamental proceeding of all. It is

intended to be the rational result of as accurate a knowledge as can be gained of the disease which has to be remedied, and of the means at our command of doing so. How this choice is to be made will be discussed under General Therapeutics in the third part of the work.

Idiosyncrasy.—Before finally deciding, however, on certain drugs, idiosyncracy must not be forgotten; that is, the peculiar susceptibility of some individuals to the action of particular medicines, such as opium, mercury, quinine, essential oils, and ipecacuanha. In almost every instance such idiosyncrasy means *increased* susceptibility; unpleasant or even dangerous results following an ordinary or even minute dose. It is well, therefore, before ordering such drugs, to enquire whether the patient has taken them previously, and if not, to use them cautiously at first.

2. **Selection of the preparation.**— The drug having been selected, the particular preparation of it will be selected in accordance with the considerations discussed under the head of varieties of preparations. The Pharmacopœia affords abundant choice, according to the channel by which it is to be administered. This naturally leads us to consider the

MODES OF ADMINISTRATION OF DRUGS.

(*a*) By the *skin*, or mucous membrane continuous with the skin, whether simply applied or rubbed in (liniment, ointment); painted on (pigment); worn on the skin (as a plaster); applied in a state of fine division by fumigation, with or without sweating; used as a gargle, injection, or wash; or insufflated on to a part. The effect desired is usually local only, but it may be general, many drugs being absorbed by the skin.

(*b*) By the *mouth*, to act locally on the alimentary canal, and to be absorbed from it, especially from the stomach.

(*c*) By the *rectum* (or *vagina* in the female), in the form of enema or injection (fluid), or of a suppository (solid). Sometimes drugs cannot be administered by the mouth, either on account of some physical obstacle, repugnance on the part of the patient, or irritability of the stomach; or to spare the strength generally, and the stomach especially, in conditions of exhaustion. Again, the action desired may be a local one on the rectum and pelvic organs, *e.g.* to relieve pain, destroy worms, or soften retained fæces.

(*d*) By injection under the skin—*subcutaneous* or *hypodermic injection*, or into the tissues—*interstitial injection*: excellent methods of admitting some remedies into the system with certainty and despatch, and in small bulk.

(*e*) By *application to wounds* or diseased surfaces, as lotions, poultices, gargles, injections, collyria; or by the *endermic* method, *i.e.* by being sprinkled on a blistered surface.

(*f*) By *inhalation*, the substances being sometimes volatile, and intended either to enter the blood through the pulmonary capillaries, *e.g.* chloroform, or to act directly on the parts to which they gain access in the form of smoke, *e.g.* cigarettes, powders, etc.; sometimes medicated watery vapours, such as Vapor Conii.

(*g*) By *intravenous injection*, very rarely practised in man.

3. **The Dose.**—The Pharmacopœia indicates the limits of ordinary doses, the minimum being the smallest useful dose which it may be wise to begin with, and the maximum being the largest usually given without special reason and caution. Experience alone can teach the practitioner how far he may safely and wisely depart from these limits, to which he is in no wise tied by law. Several modifying circumstances which are to be taken into account with respect to doses must here be carefully noted.

(*a*) Many drugs have *different actions in different doses*, which must be arranged accordingly ; *e.g.* anti-monium tartaratum, alcohol, opium, and rhubarb.

(*b*) The dose must vary with the *age* of the patient, children getting but a fraction of a dose for an adult. A convenient method of calculating the doses for children under twelve, is to divide the age in years by the age in years + 12, and to use the result as the proper fraction of an adult dose. Thus, for a child of four years the dose will be $\frac{4}{4+12} = \frac{4}{16} = \frac{1}{4}$ of an adult dose ; for a child of twelve, $\frac{12}{12+12} = \frac{12}{24} = \frac{1}{2}$. Above twelve, and under twenty-one, the dose must lie between ½ and a full dose. Delicate persons and patients exhausted by disease resemble children in bearing but small doses.

(*c*) In particular *diseases* the ordinary dose may have to be modified. In disease of the kidneys, where excretion is diminished, drugs which are dis-charged by this channel, such as morphia, are retained in the blood for a longer time, *i.e.* in larger quantity at any given time after administration, and symptoms of poisoning very readily supervene. Quite a different matter is the effect of a disease in neutralising the effect of a drug given to combat it. Thus, large doses of morphia will be tolerated in severe pain, because the action of the morphia is spent in overcoming the pain. The periods of menstruation, pregnancy, and lactation also require to be considered in prescribing.

4. **Frequency.**—Medicines are ordered to be taken one or more times, according to the desired end. Thus purgatives are generally taken in a single dose ; an emetic is to be taken once, and repeated only in case vomiting is not induced; whilst tonics are generally ordered three times a day continuously.

5. **Duration.**—The period for which a drug may be given depends entirely on a variety of circumstances

which need not be discussed here. We must refer, however, to *accumulation, toleration, custom,* and *habit*. When a drug is allowed to enter the system at short intervals, for a sufficient period, more rapidly than it can be excreted, a time will obviously come when it will have *accumulated* so much in the tissues as to produce its effects in a marked degree. Powerful drugs, *e.g.* strychnia and digitalis, may thus begin to act as poisons after having been given in the same doses with benefit for weeks. On the other hand, certain drugs lose their effect when given for a length of time, from some cause still obscure, *e.g.* opium. The dose must then be steadily increased, *toleration* being said to be established by *custom*. If a patient become dependent on a drug, crave for it, and indulge in it to an unfortunate or even vicious extent, he is said to have developed a *habit* for that drug, such as the opium and alcohol habits or the habitual use of enemata.

6. Time.—The times of the day or night at which the doses must be taken are of the first importance; and speaking generally, it may be said that every advantage must be taken in this respect of the natural tendency which it is desired to assist or stimulate by the drug. Thus, drugs which induce sleep are naturally given at bedtime; alkaline stomachics before meals; saline purgatives early in the morning. The time required by the drug to act must also be calculated, especially in the case of the different purgatives.

7. Combinations: Chemical and Physiological Incompatibles.—In most instances more than one drug has to be given at the same time, and the practitioner finds that he must combine them in a single prescription, whether, for instance, pill, powder, or liniment. Successful combination is at once the most important and difficult part of the art of prescribing. Whilst it affords the prescriber an opportunity of applying the whole of his knowledge of

drugs and their action, it cannot be accomplished without a thorough acquaintance with the physical, chemical, and physiological properties of the ingredients of the proposed compound. The mere appearance, taste, and flavour of a mixture are important points to be considered in ordering it. The chemical reactions which may occur between the constituents must be constantly kept in view. The prescriber may either intend the constituents to remain chemically unchanged, or arrange for the decomposition of one or more of them, and the production of a new substance. Drugs which decompose each other are said to be *chemically incompatible* in the widest sense; but the use of the term is commonly restricted to instances in which the result is an unexpected, inelegant, useless, or dangerous compound. Thus, if it be desired to give a patient chlorate of potash and hydrochloric acid, we say that the undiluted acid is incompatible with the salt, because chlorine is produced by their combination; but if it be intended to order a fresh solution of chlorine in water, and the decomposition is deliberately planned, the combination would not be considered incompatible. A list of incompatibles will be found under the "characters" of the principal drugs.

The prime consideration, however, will be the physiological effect of the combination. This is very different in different cases. Each of the constituents may be intended to produce an effect different from the others; or to have the same effect; or one or more ingredients may be introduced to modify the action of the principal, that is, to correct some unpleasant, dangerous, or otherwise undesirable influence which it happens to possess, in addition to the influence which we wish to secure. Such *correctives* are necessarily physiological antagonists, and appear, therefore, to be physiological incompatibles; but it is for this very reason that they are to be combined, because whilst they neutralise the action of each other in certain directions, they are left mutually free to affect other parts of the system. Thus, calomel combined with opium prevents it from causing constipation, whilst it does not interfere with its action on the brain ; and the opium, in turn, prevents the calomel from purging the patient, whilst it allows the mercurial to act as an alterative. Most purgative pills contain correctives or carminatives, which moderate the violence of peristalsis and prevent pain.

8. **The Prescription.**—A prescription consists of five parts : The *superscription*, consisting of a single sign, R, an abbreviation for *recipe*, "take"; the *inscription*, or body of the prescription, containing the names and quantities of the drugs ordered ; the *subscription*, or directions to the dispenser ; the *signature*, or directions to the patient, headed by *Signa ;* and, lastly, the patient's name, the date, and the prescriber's name or initials. In what may be called a classical prescription, it was customary to arrange the constituents of the inscription under four heads, viz. the *basis*, or active drug proper; the *adjuvant*, or substance intended to assist, and especially to hasten, the action of the basis ; the *corrective*, to limit or otherwise modify the same (commonly a carminative); and the *vehicle*, or *excipient*, to bring the whole into a convenient, pleasant form for administration.

To take an example :

Superscription.	R
Inscription.	Ferri et Ammoniæ Citratis, gr.v. (*basis*). Liquoris Ammoniæ Fortioris min.jss. (*adjuvant*). Spiritûs Myristicæ, min.vj. (*corrective*). Infusi Calumbæ, ad ʒi. (*vehicle* or *excipient*).
Subscription.	Misce. Mitte doses tales viij.
Signature.	Signa—Two tablespoonfuls twice a day.
Patient's name.	Practitioner's name
Date.	or initials.

It will be seen that the first three parts of the prescription are in Latin ; the signature or directions to the patient in English.   The names of the drugs or preparations are in the genitive case, the quantities standing in the accusative case, governed by *recipe :*

Recipe, Spiritûs Myristicæ, minima sex.

Take, of Spirit of Nutmeg, six minims.

A few *abbreviations* and *signs* are allowed, viz. : R for recipe ; *m.,* misce ; *S.,* signa ; $\overline{aa}$., ana (*avà*), of each ; *ft.,* fiat, make ; *q.s.,* quantum sufficit, a sufficiency ; *ad,* up to, to amount to (the full phrase being *quantum sufficit ad*) ; *c.,* cum, with ; *no.,* numerus, number ; *p.r.n.,* pro re natâ, as required, occasionally ; *rep.,* repetatur, let it be repeated ; *ss., ʃs.,* semi, or semis, a half.

The names of drugs must always be written in full wherever there can be the smallest possibility of error.   It is not only inelegant, but dangerous, to use such abbreviations as *Acid. Hydroc. Dil.,* and *Hyd. Chlor.*

The various weights and measures are expressed by characters and figures, very rarely by words, placed distinctly at the end of the line occupied by the name of each ingredient ; but if two or more consecutive ingredients are ordered in equal quantity, it is usual, instead of repeating this each time, to write it only once after the last of them, preceded by the sign $\overline{aa}$, of each.

Part I.

THE
INORGANIC MATERIA MEDICA.

GROUP I.

THE ALKALIES AND ALKALINE EARTHS.

Of the alkalies and alkaline earths, potassium, sodium, ammonium, lithium, calcium, magnesium, and barium are used in medicine. These, together with their many salts, alone constitute a large proportion of the inorganie materia medica. Whilst each of them is so important as to require separate consideration, many facts connected with their action and uses are common to the whole group, and much that is said under the head of Potassium, the first of the series, will not require to be repeated under the others.

POTASSIUM. K. 39.

The salts and preparations of potassium are most conveniently discussed in the following order ·

1. Potassæ Carbonas. — Carbonate of Potash, K_2CO_3, with about 16 per cent. of water of crystallisation.

Source.—Obtained from wood ashes by lixiviation, evaporation, and crystallisation.

Characters.—A white crystalline, very deliquescent powder, of caustic alkaline taste. 20 gr. neutralise 17 gr. of citric acid, or 18 gr. of tartaric acid.

Impurities.—Sulphates and chlorides.

Dose.—10 to 30 gr.

From Potassæ Carbonas are made:

a. **Potassæ Bicarbonas.**—Bicarbonate of Potash. $KHCO_3$.

Source.—Made from a solution of the Carbonate in distilled water, by passing a stream of carbonic acid gas through it, and purifying the crystals which form.

Characters.—Colourless right rhombic prisms, not deliquescent; of a saline, feebly alkaline taste; not corrosive. 20 gr. neutralise 14 gr. of citric acid, or 15 gr. of tartaric acid.

Dose.—10 to 40 gr.

Preparation.

Liquor Potassæ Effervescens. Potash Water.—Potassæ Bicarbonas, 30 gr.; Water, 1 pint. Dissolve, and pass into the solution as much CO_2 as it will contain under a pressure of 7 atmospheres.

Dose.—Ad libitum.

b. **Liquor Potassæ.**— Solution of Potash, KHO, (5·84 per cent.) in water.

Source.—Made from a solution of the Carbonate, by boiling with Slaked Lime and purifying. $K_2CO_3 + CaO,H_2O = 2KHO + CaCO_3$.

Characters.—A clear alkaline fluid.

Impurities.—Lime, carbonates, sulphates, and chlorides.

Dose.—15 to 60 min.

From Liquor Potassæ are made:

a. **Potassa Caustica.**—Caustic potash. KHO.

Source.—Made from Liquor Potassæ by evaporation.

Characters.—White pencils, hard but very deliquescent, alkaline and corrosive.

Impurities.—The same as of the liquor.

From Potassa Caustica is made:

Potassæ Permanganas. — Permanganate of potash. $KMnO_4$. See *Manganesium.*

Source.—Prepared from a mixture of caustic potash, chlorate of potash, and black oxide of manganese, by semi-fusing; then boiling, neutralising with dilute sulphuric acid, and purifying. (1) $KClO_3 + 6KHO + 3MnO_2 = 3K_2MnO_4 + KCl + 3H_2O$. (2) Boiling: $3K_2MnO_4 + 2H_2O = 2KMnO_4 + 4KHO + MnO_2$.

Characters.—Dark-purple slender prisms, inodorous, with a sweet astringent taste. Soluble in 16 parts of cold water. Should be prescribed in simple solution.

Impurities.—Sulphate of potash and oxide of manganese; detected by less solubility in water and volumetric test.

Dose.—1 to 2 gr.

Preparation.

Liquor Potassæ Permanganatis.—4 gr. in 1 fl.oz. of distilled water. *Dose,* 2 to 4 fl.dr.

β. **Potassii Iodidum.**—Iodide of Potassium. KI.

Source. — Obtained by dissolving Iodine in Liquor Potassæ, and evaporating—$6KHO + 3I_2 = 5KI + KIO_3 + 3H_2O$; then mixing the residue with wood charcoal, fusing, dissolving, and purifying, to convert the iodate, which was formed with the iodide, into iodide—$2KIO_3 + 6C = 2KI + 6CO$.

Characters.—Colourless opaque cubical crystals, with some odour of iodine, a saline taste, and feebly alkaline reaction; strikes blue with preparations containing starch on addition of chlorine.

Impurities.—Iodate; detected by blue colour with tartaric acid and starch. Free iodine; by starch. And the impurities of the liquor.

Dose.—2 to 10 gr., or more.

For *Preparations,* see *Iodum.*

γ. **Potassii Bromidum.** — Bromide of Potassium. KBr.

Source.—Obtained from Liquor Potassæ, Bromine, and Charcoal, by the same process as the iodide.

Characters.—Colourless cubic crystals, without odour, and of a pungent saline taste. Does not strike blue with preparations containing starch, unless it contain iodide as impurity.

Dose.—5 to 30 gr.

ε. **Potassæ Citras.**—Citrate of Potash. $K_3C_6H_5O_7$.

Source.—Made by neutralising a solution of Citric Acid with Carbonate of Potash, and evaporating. $3K_2CO_3 + 2C_6H_5O_7, H_3 = 2K_3C_6H_5O_7 + 3H_2O + 3CO_2$.

Characters.—A white deliquescent powder, of saline, feebly acid taste.

Dose.—20 to 60 gr.

d. **Potassæ Acetas.**—Acetate of Potash. $KC_2H_3O_2$.

Source.—Made by dissolving Carbonate of Potash in Acetic Acid, evaporating, and fusing the residue. $K_2CO_3 + 2(HC_2H_3O_2) = 2(KC_2H_3O_2) + H_2O + CO_2$.

Characters.—White foliaceous satiny masses, very deliquescent; neutral. The peculiar appearance of this salt is due to crystallisation after fusion.

Impurities.—The carbonate; detected by being insoluble in spirit. Excess of acid; giving acid reaction. Metallic impurities.

Dose.—10 to 60 gr.

Potassæ Acetas is used in preparing :

Tinctura Ferri Acetatis. See *Ferrum.*

e. **Potassæ Chloras.**—Chlorate of Potash. $KClO_3$.

Source.—Made by passing Chlorine gas into a mixture of Carbonate of Potash and Slaked Lime (*i.e.* caustic potash), boiling, evaporating, and separating the chloride of potassium by re-crystallisation. $6KHO + 6Cl = KClO_3 + 5KCl + 3H_2O$.

Characters. — Colourless rhomboidal crystalline plates, with a cool, sharp, saline taste. Explodes when rubbed with sulphur. Soluble in 16 parts of cold water.

Impurities.—Chloride of calcium, and lime.

Dose.—10 to 30 gr.

Preparation.

Trochisci Potassæ Chloratis.—5 gr. in each.

f. **Potassa Sulphurata.**—Sulphurated Potash.

Source.—Made by fusing together Carbonate of Potash and Sublimed Sulphur.

Characters.—Solid greenish masses, liver-brown when recently broken, alkaline and acrid to the taste; a mixture of sulphide, sulphate, sulphite, and hyposulphite.

Dose.—3 to 8 gr.

Preparation.

Unguentum Potassæ Sulphuratæ.—1 in 15½.

2. Potassæ Tartras Acida.—Acid Tartrate of Potash, Bitartrate of Potash, Cream of Tartar. $KHC_4H_4O_6$.

Source.—Prepared from argol, the deposit in wine-casks.

Characters.—A white gritty powder, or fragments of cakes, of a pleasant acid taste ; not deliquescent. Soluble in 200 parts of cold water.

Dose.—20 to 60 gr. as a diuretic and refrigerant; 2 to 8 dr. as a purgative.

Acid Tartrate of Potash is an important ingredient of :

Confectio Sulphuris (1 in 9); and Pulvis Jalapæ Compositus (9 in 15). It is also used in preparing various other drugs.

From this salt is derived :

Potassæ Tartras.—Tartrate of Potash. $K_2C_4H_4O_6$.
Source.—Made by adding Acid Tartrate of Potash to a solution of Carbonate of Potash, and crystallising. $2KHC_4H_4O_6 + K_2CO_3 = 2K_2C_4H_4O_6 + CO_2 + H_2O$.
Characters.— Small colourless deliquescent prisms. 10 parts are dissolved in 8 parts of water.
Impurities.—Acid tartrate; detected by insolubility. Carbonates; by quantitative test.
Dose.—20 to 60 gr. as a diuretic and antacid; 2 to 4 dr. as a purgative.

3. Potassæ Nitras.— Nitrate of Potash, Nitre, Saltpetre. KNO_3.

Source.—Found native, and purified by crystallisation.
Characters.—Striated colourless prisms, of a peculiar saline taste. Solubility, 1 in 4 of water.
Dose.—10 to 30 gr.

From Potassæ Nitras is made :

Potassæ Sulphas.—Sulphate of Potash. K_2SO_4.
Source.—Prepared from Nitrate of Potash and Sulphuric Acid, which yield the acid sulphate—$KNO_3 + H_2SO_4 = KHSO_4 + HNO_3$; then adding Carbonate of Potash—$2KHSO_4 + K_2CO_3 = 2K_2SO_4 + H_2O + CO_2$.
Characters. — Colourless hard six - sided prisms, terminated by six-sided pyramids. Solubility, 1 in 10 of water; not soluble in spirit.
Impurities.—Other sulphates, and chlorides.
Dose.—15 to 60 gr.

Potassæ Sulphas is contained in :

Pulvis Ipecacuanhæ Compositus. — 8 in 10; Pilula Colocynthidis Composita; and Pilula Colocynthidis cum Hyoscyamo.

ACTION AND USES.

1. IMMEDIATE LOCAL ACTION AND USES.

Externally.—Potash, in the form of potassa caustica, is a powerful irritant and caustic, absorbing water from the part to which it is applied, and converting it into a moist, grey slough. It is used to destroy morbid growths, to form issues, and to stimulate ulcers. Solutions of liquor potassæ or the carbonates neutralise caustic acids on the skin. Hot dilute solutions relieve the pains of rheumatism and gout when used as local baths or fomentations to the affected joints. Weak compounds of potash with olive oil constitute soft soaps, which also have antacid and cleansing properties.

Internally.—Potash and its salts have an alkaline action, and are employed as antidotes to the caustic acids; but the use of the carbonates for this purpose ought, if possible, to be avoided, on account of the great development of carbonic acid. In the mouth, potash checks for a moment the secretion of saliva. Reaching the stomach, it partly neutralises the contents; and liquor potassæ effervescens will relieve acidity due to excess of gastric juice, or to the decomposition attending indigestion. Of much greater importance is the stomachic action of potash given shortly before meals, when, as a dilute alkali, it is a natural stimulant to the gastric follicles, increasing the flow of the juice, and is a sedative to the nerves. Liquor potassæ and bicarbonate may be used for this purpose in dyspepsia, especially when there is much pain and tendency to sickness, or when the further action of potash on the system is desired, as in gouty, rheumatic, and calculous subjects; but soda is more commonly employed. Large doses of the bicarbonate are apt to irritate the stomach.

Some valuable saline purgatives belong to the potash group, notably the acid tartrate, tartrate, and sulphate. The rationale of the action of saline purgatives is discussed in Part III. In dropsy from any cause, especially ascites from liver disease, the acid tartrate, in the form of Pulvis Jalapæ Compositus, of an electuary with honey, or of a lemonade, may be used to remove the water by the bowels, its hydragogue effect being assisted by its action as a diuretic.

2. ACTION ON THE BLOOD AND ITS USES.

Potash is freely absorbed into the blood in the form of salts, and there acts both on the plasma and red corpuscles, increasing the natural alkalinity of the former, and improving the quality and increasing the number of the latter when judiciously combined with iron. As an **alkaliniser** of the plasma, although exceedingly transitory in its action, being very rapidly excreted, potash is a valuable remedy in gout, where it combines with the excess of uric acid in the blood and facilitates its excretion. The carbonates, citrates, and tartrates of potash in various forms, and the waters of such spas as Baden-Baden, Wiesbaden, Vichy, Carlsbad, and Aix-la-Chapelle, which contain definite though small quantities of potassium salts, are extensively used for the treatment of acute and chronic gout. The salts of the vegetable acids, or the effervescing carbonates, are the best preparations for prolonged use. In acute rheumatism the bicarbonate, citrate, tartrate, and acetate are successfully employed to increase the alkalinity of the blood. For restoration of the red corpuscles in anæmia by the increase of their potash element, the carbonate is the best **hæmatinic**, either as contained in the Mistura Ferri Composita, or given as a pill with sulphate of iron (Blaud's Pill).

An indirect action of potash on the blood must here be carefully noted. We shall see hereafter that citric, tartaric, and acetic acids, given internally, are partially oxydised in the blood. The completeness of the combustion, and of the important influences which the change exerts on the blood and kidneys, depends upon the combination of the vegetable acid with an alkali. Citric acid, *e.g.*, is excreted mostly unchanged in the urine, but citrate of potash is entirely, or almost entirely, thrown out as the carbonate.

3. SPECIFIC ACTION AND USES.

Potash depresses the **muscular, nervous,** and **cardiac** tissues; and the point of interest in this connection is, that when given for other purposes it must be used with caution. The danger of "potash poisoning" is, however, exaggerated, for the drug passes so quickly through the system, that it cannot well produce a deleterious effect on the tissues, unless given for a very long time, or in disease of the excreting organs, especially the kidneys. Excessive single doses are generally rejected at once by vomiting.

4. REMOTE LOCAL ACTION AND USES.

Potash is excreted almost entirely by the kidneys; to a much less extent by the skin, respiratory passages, stomach, liver, biliary passages, and bowels. In other words, it passes out in the fluids of all the secretory surfaces, and in doing so it stimulates the cells to increased activity.

The diuretic effect of several potash salts, referable to their influence upon the renal epithelium, is the most important of all; and the acetate, acid tartrate, citrate and tartrate, carbonate, bicarbonate, and sulphate are used for this purpose in the order named. These saline diuretics are given chiefly in renal dropsy, where it is desirable to increase the functional activity of the renal epithelium, and thus the secretion both of water and urea, whilst the vessels remain undisturbed. They are also suitable diuretics in feverish conditions. In cardiac dropsy they are less beneficial, as they diminish rather than increase the force of the circulation; but in an occasional full dose they are useful adjuvants, even in this condition, to other classes of diuretics, such as digitalis and scoparium, to wash out the tubules. Nitrate of potash is a powerful diuretic, belonging partly to a different class, the local vascular stimulants. It is more suitably employed as a diuretic in feverish conditions, and to remove inflammatory effusions into the pleura and pericardium, and must be given with caution in renal disease.

As **alkalinisers of** the **urine**, the carbonate, bicarbonate, and the vegetable salts of potash are extensively used in uric acid gravel, acute and chronic gout, and acute rheumatism, the latter being preferred because less irritant. In uric acid calculus of the kidney or bladder, these salts have been successfully employed to cause actual solution of the concretions.

The **diaphoretic** effect of potash salts is not marked, the citrate and the nitrate alone being used for this purpose, and that only in mild feverish attacks.

The bronchial secretions may be increased and rendered less tenacious in inflammation and dry catarrh of the tubes by the potash salts, which are thus saline expectorants, the iodide in particular being useful for this purpose.

Gastric catarrh, especially in gouty subjects, is benefited by the milder salts of potash beyond their immediate local effect; but the mineral waters which appear to act in this way, such as those of Vals, Vichy, and Carlsbad, owe their efficiency much more to soda. The same remarks apply to catarrh of the *biliary* passages and tendency to gall-stones.

The action of potash on the *intestinal glands* constitutes it a remote as well as an immediate purgative.

ACTION AND USES OF THE DIFFERENT SALTS OF POTASH.

On reviewing what has been said respecting potash, we find that the chief actions and uses of its different salts may thus be briefly represented: *Potassa Caustica*—caustic. *Liquor Potassæ*—antacid and stomachic. *Potassæ Bicarbonas, Carbonas,* and *Citras*—antacid stomachics, alkalinisers of blood and urine, mild diuretics, very mild diaphoretics, saline expectorants, biliary stimulants. *Potassæ Tartras, Tartras Acida*, and *Acetas* —the same, but more powerful diuretics; also saline purgatives. *Potassæ Sulphas*—chiefly purgative. *Potassæ Nitras* —excreted unchanged in the urine; is a double diuretic, and probably in this way only a mild febrifuge. *Potassæ Chloras*—excreted unchanged in all the secretions, including the saliva; and is much used in inflamed, ulcerative, and aphthous states of the mouth. The remaining salts of potassium contain, in combination with the alkali, an element or acid possessing such distinctly specific actions that the total effect is but in a minor degree referable to the potash. Such are the arsenite, bromide, iodide, and permanganate, and sulphurated potash. These will, therefore, be discussed under the head of their other constituents.

SODIUM. Na. 23.

The following are the officinal salts and preparations of sodium, arranged according to their source:

Source.—Made from the ashes of marine plants, or from chloride of sodium by chemical decomposition.

Characters.—Transparent colourless laminar rhombic crystals, efflorescent, with a harsh alkaline taste, and alkaline reaction. 20 gr. neutralise 9·7 gr. of citric acid, or 10½ gr. of tartaric acid.

Impurities.—Sulphates and chlorides.

Dose.—5 to 30 gr.

From Sodæ Carbonas are made:

a. **Sodæ Carbonas Exsiccata.**—Dried Carbonate of Soda. Na_2CO_3. A dry white powder, made from Carbonate of Soda by drying. 53 gr. = 143 gr. of the crystallised salt. *Dose,* 3 to 10 gr.

b. **Sodæ Bicarbonas.**—Bicarbonate of Soda. $NaHCO_3$.

Source.—Prepared from a mixture of the Carbonate and Dried Carbonate by passing a stream of carbonic acid gas through them; $Na_2CO_3 + H_2O + CO_2$ $2(NaHCO_3)$.

Characters.—A white powder, or small opaque irregular scales, of a not unpleasant saline taste. 20 gr. neutralise 16·7 gr. of citric acid, or 17·8 gr. of tartaric acid. 1 part soluble in 10 of water.

Impurities.—Carbonate and its impurities.
Dose.—10 to 60 gr.

Preparations.

α. **Liquor Sodæ Effervescens.**—Soda Water. Made like potash water. 30 gr. in 1 pint. *Dose,* Ad libitum.

β. **Trochischi Sodæ Bicarbonatis.**—5 gr. in each. *Dose,* 1 to 6.

γ. **Sodæ Citro-Tartras Effervescens.**—Effervescent Citro-tartrate of Soda. White granules, deliquescent; made by heating the Bicarbonate with Citric and Tartaric acids, stirring assiduously. *Dose,* 60 to 120 gr.

c. **Liquor Sodæ.**—Solution of Soda. NaHO (4·1 per cent.) in water.

Source.—Prepared by boiling a solution of Carbonate of Soda with Slaked Lime. $Na_2CO_3 + CaH_2O_2 = 2NaHO + CaCO_3$.

Characters.—A colourless alkaline liquid.
Impurities.—Lime, carbonates, sulphates, chlorides.
Dose.—(Rarely given) 10 min. to 1 fl.dr.

From Liquor Sodæ are made :

α. **Soda Caustica.**—Caustic Soda. NaHO.
Source.—Made by boiling down Liquor Sodæ.
Characters.—Hard greyish-white fragments, slightly deliquescent, very alkaline.
Impurities.—Same as of liquor.

β. **Sodæ Valerianas.** See *Valerianæ Radix.*

d. **Sodæ Acetas.**—Acetate of Soda. $NaC_2H_3O_2$. $3H_2O$.

Source.—Made by acting on Carbonate of Soda by Acetic Acid.

Characters.—Transparent colourless crystals. Used only to make ferri arsenias, ferri phosphas, and syrupus ferri phosphatis.

e. **Soda Tartarata.**—Tartarated Soda. NaK. $C_4H_4O_6.4H_2O$. Tartrate of soda and potash. Rochelle salt.

Source.—Prepared by boiling Acid Tartrate of Potash in a solution of Carbonate of Soda, and crystallising; $Na_2CO_3 + 2KHC_4H_4O_6 = 2NaK.C_4H_4O_6 + H_2O + CO_2$.

Characters. — Colourless transparent right rhombic

prisms, tasting like common salt. Solubility, 1 in 2 of water.

Impurities.—Acid tartrate of potash.

Dose.—¼ to ½ oz.

f. **Sodæ Sulphas.**—Sulphate of Soda. Glauber's salt. $Na_2SO_4.10H_2O$.

Source.—Prepared by adding Carbonate of Soda to the acid sulphate left in the manufacture of hydrochloric acid. $Na_2CO_3 + 2NaHSO_4 = 2Na_2SO_4 + CO_2 + H_2O$.

Characters.—Colourless transparent oblique rhombic prisms, efflorescent, with a bitter salt taste. 1 part soluble in 3 of water.

Impurities.—Salts of ammonium and iron.

Dose.—¼ to 1 oz.

g. **Sodæ Phosphas.**—Phosphate of Soda. $Na_2HPO_4.12H_2O$.

Source.—Prepared by adding a solution of Carbonate of Soda to the acid product of the action of sulphuric acid on bone-ash, and crystallising. (1) $Ca_32PO_4 + 2H_2SO_4 = CaH_42PO_4 + 2CaSO_4$. (2) $CaH_42PO_4 + Na_2CO_3 = Na_2HPO_4 + CaHPO_4 + H_2O + CO_2$.

Characters.—Colourless transparent rhombic prisms, efflorescent, tasting like common salt. 1 part soluble in 5 of water. Used to make ferri phosphas and syrupus ferri phosphatis.

Impurity.—Phosphate of lime.

Dose.—¼ to 1 oz.

h. **Sodæ Hypophosphis.**—Hypophosphite of Soda. $NaPH_2O_2$.

Source.—Prepared by adding Carbonate of Soda to a solution of Hypophosphite of Lime, and evaporating the solution. $Ca2PH_2O_2 + Na_2CO_3 = 2NaPH_2O_2 + CaCO_3$.

Characters. — A white granular deliquescent salt, with a bitter nauseous taste. 1 part soluble in 2 of water.

Dose.—5 to 10 gr.

i. **Sodæ Arsenias.**—Arseniate of Soda. $Na_2HAsO_4.7H_2O$.

Source.—Prepared by fusing Carbonate of Soda and Nitrate of Soda with Arsenious Acid, dissolving out and crystallising.

Characters.—Colourless transparent prisms. 1 part soluble in 2 of water.

Dose.—1/16 to ⅛ gr.

Preparation.

Liquor Sodæ Arseniatis.—4 gr. dried to 1 oz. water. *Dose,* 5 to 10 min.

j. Sodæ Chloratæ Liquor.—Solution of Chlorinated Soda.

Source.—Made by passing a stream of Chlorine gas through a solution of Carbonate of Soda.

Characters. — A colourless liquid, with an odour of chlorine and an astringent taste; alkaline. A mixed solution of hypochlorite of soda, chloride of sodium, and carbonate of soda.

Dose.—10 to 20 min.

Preparation.

Cataplasma Sodæ Chloratæ.

2. Sodii Chloridum.—Chloride of Sodium. Common salt. NaCl.

Source.—Native.

Characters.—Small white crystalline grains, or transparent cubic crystals, free from moisture, with purely saline taste. 1 part soluble in $2\frac{3}{4}$ parts of water.

Dose.—10 to 240 gr.

Sodii Chloridum is used in making :

Acidum Hydrochloricum, Hydrargyri Perchloridum, and Hydrargyri Subchloridum.

3. Sodæ Nitras.—Nitrate of Soda. $NaNO_3$.

Source.—Native; purified by crystallisation from water.

Characters. — Colourless obtuse rhombohedral crystals, with a cooling saline taste.

Impurities.—Sulphates and chlorides.

Sodæ Nitras is used in making :

Sodæ Arsenias.

4. Sodæ Biboras. — Biborate of Soda. Borax. $Na_2B_4O_7.10H_2O$.

Source.—Native.

Characters. — Transparent colourless crystals, slightly efflorescent, weakly alkaline. 1 part soluble in 22 parts of cold water, or in 1 part of glycerine.

Dose.—5 to 40 gr.

Preparations.

a. Glycerinium Boracis.—1 to 4.

b. Mel Boracis.—1 in 8.

ACTION AND USES.

1. IMMEDIATE LOCAL ACTION AND USES.

Externally, soda possesses an action similar to that of potash, but is much less frequently used as a caustic. Solutions of the carbonates may be employed to neutralise caustic acids; in eczema and itching disorders of the skin ; and in extensive burns.   Soda compounds with olive oil constitute hard soap.

Internally.—Soda closely resembles potash in its action on the alimentary canal, but is more powerful because much less diffusible, *i.e.* more slowly absorbed.   It is antacid to the contents of the stomach, relieving acidity due to indigestion, in the form either of the bicarbonates, soda-water, the officinal lozenges, or as a mixture with sal-volatile and an essential oil, such as peppermint, given after meals.   As a stomachic, stimulating the flow of the gastric juice, bicarbonate of soda is more commonly given than the other alkalies, in doses of gr. 8 to gr. 15, shortly before meals.   Part of the salt at the same time becomes converted into the chloride, which assists the digestion of albumen.   The alkali also liquefies tenacious mucus, and thus prevents decomposition, the juice reaching the food more freely.   Common salt is a safe and available emetic.

The salts of soda, being much less diffusible than those of potash, pass on into the small intestine.   Here the sulphate and phosphate of soda and tartarated soda (Rochelle salt) act as saline purgatives.   The sulphate, which is a constituent of several natural purgative waters, including Carlsbad, Friedrichshall, Hunyàdi János, and Bilin, is the most powerful of these, producing an abundant watery evacuation.   It is used as a hydragogue in dropsies, especially in ascites from liver disease, in congestion of the portal system, and as a habitual purgative. The phosphate is a milder, but sufficiently active, purgative, less unpleasant to the palate; it is often given to children.   Soda tartarata, the purgative basis of the Seidlitz powder, is familiar as a milder intestinal stimulant, of use in completing the effect of purgative pills.   The chloride is an anthelmintic.

2. ACTION ON THE BLOOD AND ITS USES.

The salts of soda are slowly absorbed into the blood, and slowly excreted from it, remaining in it chiefly as the bicarbonate and phosphate.   Taken, as they constantly are, in food, these salts are the chief sources of the natural alkalinity of the liquor sanguinis, which may be increased by their medicinal exhibition as well as by the tartrate, Rochelle salt, and even the sulphate.   This effect of soda as an alkaliniser of the blood is taken advantage of in the cases referred to under potash, namely, gout and rheumatism, only less frequently; for although soda is less depressing, as we shall see, than potash, and more easily borne on the stomach, the slowness of its entrance into the blood, and its tendency to pass off by the bowels when the dose is increased, more than counteract these advantages. When a prolonged and moderate alkaline influence is desired, especially in dyspepsia with a tendency to constipation, soda is manifestly to be preferred.

3. SPECIFIC ACTION.

In medicinal doses, the salts of soda have **no specific influence** on any organ. This circumstance, which at first sight appears incredible, is due to the fact that the whole organism is saturated with soda, which participates in many of the ordinary tissue changes; that soda is admitted in large quantities by the food (especially vegetables and fruits); and that the moderate amount contained in medicinal doses does not appreciably affect metabolism. In this respect soda differs remarkably from potash, and is therefore said to produce none of the depressing effects of that drug. As we have just seen, advantage is taken of this negative action of soda in its therapeutical applications.

4. REMOTE LOCAL ACTION AND USES.

Soda is excreted by all the mucous surfaces, by the kidneys, by the liver, and possibly by the skin; and in passing through the various epithelial structures, it increases their activity, whilst it modifies the amount, composition, and reaction of their secretions. The action of the different salts naturally varies to a considerable extent, some affecting one organ more, some another. The sulphate and the phosphate of soda are, as we have seen, hydragogue purgatives by virtue of their *immediate* local action; but they are also stimulants of the intestinal glands, and are constantly being absorbed and excreted, re-absorbed and re-excreted, in their course along the bowel. (*See* Part III.) Both are also true hepatic stimulants or direct cholagogues; the phosphate more so than the sulphate. The value of these salts in hepatic and intestinal disorders, which has been already referred to, is therefore partly referable to their effect in increasing the bile. Soda tartarata has a similar but feebler action.

On the *kidneys* soda acts less powerfully as a diuretic than potash, increasing the water and the solid constituents, and diminishing or neutralising the acidity of the urine. The bicarbonate is the most useful salt of soda for this purpose; the acetate and nitrate, whilst also diuretic, are so inferior to the acetate and nitrate of potash, that they are very seldom employed. The tartarated soda may be usefully combined with other **alkalinisers of the urine**, as in the ordinary Seidlitz powder ; and the effervescing citro-tartrate of soda is a convenient modification of much the same drugs. The use of these alkalinisers of the urine has been explained already.

The secretions of the *bronchi* are increased by soda ; that is, the sputa become more abundant and more liquid, and are more easily expelled by cough. The bicarbonate and the chloride are specially expectorant, and are indicated in the early stages of bronchitis, and in recurrent slight bronchial catarrh, when secretion is deficient and cough harassing. The effects of soda on the stomach, blood, and urine add much to its usefulness in such cases. The stimulant effect of soda salts on ciliary action may also account in part for its expectorant action.

When a comprehensive view is taken of the action and uses of the salts of soda—locally in the alimentary canal, in the blood, in the tissues, and in the organs and passages where it is excreted from the body, it is found to be peculiarly indicated in a condition of system which may be called the " gouty," the "rheumatic," "acidity," or " chronic derangement of the liver," and which is specially characterised, amongst other symptoms, by catarrhs, or discharges from the mucous membranes, interfering with the functions of the part ; by imperfect biliary activity and constipation ; and by scanty, high-coloured, very acid urine. In such a condition great benefit may be derived from a course of alkaline waters. If the stomach be the principal seat of catarrh, *i.e.* if chronic indigestion be urgent, the more purely carbonated alkaline waters should be selected, such as those of Vichy, Bilin, and Ems. If the derangement chiefly involve the liver and intestines, the sulphated and salt (NaCl) waters will be more suitable, such as Carlsbad, Kissingen, Wiesbaden, and Marienbad. For chronic catarrh of the bladder and urinary passages, Ems, Vichy, Wildungen, and Carlsbad are indicated.

5. ACTION AND USES OF THE DIFFERENT SODA SALTS.

The action and uses of the preparations of soda may be summarised as follows, and the special action of some of the salts particularly noticed : *Soda Caustica* and *Liquor Sodæ* are for external use, but very rarely employed. *Sodæ Carbonas* and *Bicarbonas* (the former rarely, the latter almost invariably used) possess the action and uses of soda in general upon all parts. *Sodæ Citro-tartras* is like the carbonates, but milder. *Soda*

Tartarata is like the carbonates, but purgative; and more rapidly and distinctly diuretic and alkalinising, by virtue of the potash it contains. *Sodæ Acetas* and *Sodæ Nitras* are used pharmaceutically only. *Sodæ Sulphas* and *Sodæ Phosphas* are chiefly hydragogue purgatives and cholagogues, the former acting more on the bowels, the latter more on the liver. *Sodii Chloridum* is in large doses a free and safe emetic; an anthelmintic as enema; it possesses otherwise the ordinary action of soda, and is greatly used for this purpose as the waters of Homburg, Wiesbaden, Kissingen, and Baden-Baden, and as sea-water. The remaining salts of soda possess peculiar properties by virtue of their second constituent, and are described elsewhere—*Sodæ Arsenias* under Arsenic; *Sodæ Chloratæ Liquor*, under Chlorine; *Sodæ Hypophosphis* under Phosphorus; *Sodæ Biboras* under Acidum Boricum; *Sodæ Valerianas* under Valerianæ Radix.

AMMONIUM. NH_4.   18.

The following are the officinal salts and preparations of ammonium, arranged according to their source:

1. Ammonii Chloridum.—Chloride of Ammonium. Muriate of Ammonia. Sal Ammoniac. NH_4Cl.

Source.—Made by neutralising Ammoniacal Gas Liquor with Hydrochloric Acid, and evaporating. $NH_4HO + HCl = NH_4Cl + H_2O$.

Characters.—Colourless translucent fibrous masses, inodorous and tough. Solubility, 1 in 4 of water; soluble in spirit.

Impurities.—Iron and lead.

Dose.—5 to 20 gr.

From Ammonii Chloridum are made

a. **Liquor Ammoniæ Fortior.**— Strong solution of Ammonia. NH_3 (32·5 per cent.) in water.

Source.—Made by heating Chloride of Ammonium with Slaked Lime, and collecting the gaseous product in water. $2NH_4Cl + CaH_2O_2 = 2NH_3 + CaCl_2 + 2H_2O$.

Characters.—A colourless liquid with a very pungent characteristic odour, and strong alkaline reaction.

Impurities. — Ammonium chloride, sulphide, and sulphate; lime, and metals. Detected by usual tests.

Preparations.

a. Linimentum Camphoræ Compositum.—Compound Liniment of Camphor. Strong solution of Ammonia, Camphor, Rectified Spirit, and Oil of Lavender; 1 in 4½.

β. **Liquor Ammoniæ Citratis.**—Solution of Citrate of Ammonia. $3NH_4.C_6H_5O_7$ dissolved in water. Made by adding Strong Solution of Ammonia to a solution of Citric Acid.

Dose.—2 to 6 fl.dr.

γ. **Spiritus Ammoniæ Aromaticus.**—Aromatic Spirit of Ammonia. Sal-volatile. Made by distilling a mixture of Strong Solution of Ammonia, Carbonate of Ammonia, Volatile Oil of Nutmeg, Oil of Lemon, Spirit, and Water.

Dose.—½ to 1 fl.dr.

Spiritus Ammoniæ Aromaticus is used in preparing Tinctura Guaiaci Ammoniata, and Tinctura Valerianæ Ammoniata.

δ. **Spiritus Ammoniæ Fœtidus.**—Fetid Spirit of Ammonia. Made by adding Strong Solution of Ammonia to an extract made from Assafœtida macerated in spirit.

Dose.—½ to 1 fl dr.

ε. Tinctura Opii Ammoniata. See *Opium.*

From Liquor Ammoniæ Fortior are made :

ζ. **Ammoniæ Phosphas.**—Phosphate of Ammonia. $(NH_4)_2HPO_4$.

Source.—Made by adding Strong Solution of Ammonia to Diluted Phosphoric Acid, evaporating and crystallising.

Characters.—Transparent colourless prisms, becoming opaque by exposure. Solubility, 1 in 2 of water.

Dose.—5 to 20 gr.

η. **Liquor Ammoniæ.**—Solution of Ammonia. NH_3 (10 per cent.) dissolved in water.

Source.—Made by mixing one part of Strong Solution of Ammonia with two parts of Distilled Water.

Dose.—10 to 20 min. diluted.

Preparation.

Linimentum Ammoniæ.—Liniment of Ammonia. Solution of Ammonia, 1 ; Olive Oil, 3.

From Liquor Ammoniæ are made.

i. **Ammoniæ Benzoas.**—Benzoate of Ammonia. $NH_4.C_7H_5O_2$.

Source.—Made by dissolving Benzoic Acid in Solution of Ammonia, evaporating, and crystallising.

Characters.—Colourless laminar crystals, with a characteristic odour. Solubility, 1 in 5 of water.

Dose.—10 to 20 gr.

ii. **Ammoniæ Nitras.**—Nitrate of Ammonia. NH_4NO_3.

Source.—Made by neutralising Diluted Nitric Acid with Solution of Ammonia (or Carbonate of Ammonia), crystallising, and fusing.

Characters.—A white deliquescent salt, in confused crystalline masses, with a bitter acrid taste.

Used only for making nitrous oxide gas (N_2O).

b. **Ammoniæ Carbonas.**— Carbonate of Ammonia. $N_4H_{16}C_3O_8$.

Source.—Made by subliming a mixture of Chloride of Ammonium (or Sulphate of Ammonia), with Carbonate of Lime. $6NH_4Cl + 3CaCO_3 = N_4H_{16}C_3O_8 + 3CaCl_2 + 2NH_3 + H_2O$.

Characters.—Translucent crystalline masses, volatile and pungent to the nose ; alkaline. Solubility, 1 in 4 of water. 20 gr. neutralise $23\frac{1}{2}$ gr. citric acid, or $25\frac{1}{2}$ gr. tartaric acid ; 15 gr. neutralise 17 gr. citric acid, or one tablespoonful of lemon juice.

Impurities.—Sulphates and chlorides.

Dose.—3 to 10 gr.

From Ammoniæ Carbonas are made :

α. Spiritus Ammoniæ Aromaticus. *See* page 43.

β. **Liquor Ammoniæ Acetatis.** — Solution of Acetate of Ammonia. $NH_4.C_2H_3O_2$ dissolved in water.

Source.—Made by neutralising Carbonate of Ammonia by Acetic Acid, and adding water.

Dose.—2 to 6 fl.dr.

Ammonii Chloridum is used in making Liquor Hydrargyri Perchloridi.

2. Ammonii Bromidum.— Bromide of Ammonium. NH_4Br.

Characters.— Colourless crystals, which become slightly

yellow by exposure to the air, and have a pungent saline taste. Solubility, 1 in 1½ of water.

Impurities.—Iodides; free bromine. Detected by colour.

Dose.—2 to 20 gr.

ACTION AND USES.

1. IMMEDIATE LOCAL ACTION AND USES.

Externally applied, ammonia is a stimulant to the nerves and other structures, causing a sensation of pain and burning, and reddening the part by dilating the vessels. If the application be prolonged and the vapour confined, blistering may result; but dilute preparations produce only a rubefacient effect and a sense of heat. It is used in the form of liniment to stimulate the circulation in a part, either for the purpose of increasing the local nutrition (for instance, in stiffness or other chronic conditions of joints), or as a counter-irritant (*see* Part III.) in diseases of deeper parts, *e.g.* on the surface of the chest in bronchitis. Ammonia is not to be used as a caustic; and vesication by it is better avoided. In serpent's bite, the application of ammonia to the wound has occasionally saved life.

Internally.—Admitted into the *nose*, ammonia itself, or the vapour of the carbonate ("smelling salts"), is a powerful general stimulant, instantly causing a pungent sensation, sneezing and other disturbances of the respiration, acceleration of the pulse, and watery secretion from the parts including the conjunctiva. It is accordingly used as a means of resuscitating consciousness, the action of the heart, and respiration, in cases of failure of the circulation, such as fainting, or of asphyxia.

In the *stomach*, ammonia produces the same effects as on the skin. A full dose (30 gr. of the carbonate well diluted) is an emetic, which is best used in croup and bronchitis. Smaller doses cause a sense of warmth at the epigastrium, and act as carminatives, sal-volatile being chiefly used for this purpose. In common with soda and potash, it has an antacid effect on the contents of the stomach, and may be given after meals in dyspepsia. Like these, also, it acts as a natural stimulant to the gastric juice before meals, and sal-volatile is therefore a common ingredient of alkaline stomachic mixtures. On the bowels, ammonia appears to have no *local* action.

2. ACTION ON THE BLOOD, AND ITS USES.

Ammonia is absorbed into the blood, and is there fixed; increasing, possibly, the alkalinity of the plasma, and diminishing

the tendency to coagulation.   The phosphate is believed to
be useful in gout, by keeping uric acid in solution.

3. SPECIFIC ACTION AND USES.

Although its specific action is still imperfectly known, ammonia certainly appears to stimulate the central nervous system
generally, the respiratory centre, and the heart ; that is, to be a
general stimulant.   It is much given in neuralgia (as the
chloride), and in exhausted states of the vital powers, especially
if respiration and circulation threaten to fail, as in typhoid
fever complicated with pneumonia, in the bronchitis of old or
weakly subjects, and in ordinary acute pneumonia with increasing feebleness of the heart.   In this way also it is useful in
serpent's bite, and is given internally in water, or hypodermically (10 to 20 minims) whilst it is applied to the wound.   The
phosphate directly increases the amount of bile, etc. ; chloride of
ammonium decidedly increases the production of urea, partly,
at least, by its own decomposition in the liver.

4. REMOTE LOCAL ACTION, AND USES.

Ammonia is excreted by the kidneys and mucous membranes, especially the respiratory tract ; not, however, as ammonia, but as some other nitrogen compound.   Thus, instead
of diminishing, it actually increases the acidity of the urine,
whilst the amount of urea and uric acid also rises, as well as
the volume of the secretion.   The chloride of ammonium possesses these important powers most fully, the acetate less fully.
They are employed as diuretics in dropsies and feverish states
of the system.

The *bronchial* secretion is distinctly increased, and rendered more liquid and easily raised, by the carbonate and
chloride of ammonium.   These salts prove of great service
as expectorants in the treatment of bronchitis when the
secretion is scanty and thick, or the patient feeble ; the
accompanying stimulation of the respiratory centre increasing
the coughing or expectorant power, whilst the heart is also
sustained.

The mucous secretion of the *stomach* is affected by ammonia
as by the other alkalies, and the chloride is sometimes used in
chronic dyspepsia.   Ammonia remotely stimulates the intestines, and will cause diarrhœa if given in large doses.

On the *skin* the acetate of ammonia acts as a well-marked
remote stimulant, and as the Liquor is one of our most common
diaphoretics.   The chloride also possesses the same property,
but to a less degree.

5. ACTION AND USES OF THE DIFFERENT SALTS OF AMMONIA.

These may be thus summarised: *Liquor Ammoniæ Fortior* and *Liquor Ammoniæ* are used as stimulants and vesicants, the former externally only. *Ammoniæ Carbonas*—a volatile stimulant, emetic, and double expectorant (through the nerves and secretions). *Ammonii Chloridum*—a local refrigerant, a gastric, intestinal, and hepatic stimulant, nervous stimulant, diuretic double expectorant, and diaphoretic (hence called an "alterative"). *Liquor Ammoniæ Acetatis*—diaphoretic and diuretic (febrifuge), and nervous stimulant. *Liquor Ammoniæ Citratis* — diuretic and diaphoretic. *Spiritus Ammoniæ Aromaticus* — agreeable and powerful carminative, antacid, and general stimulant. *Ammoniæ Phosphas*—direct cholagogue, possibly an alkaliniser of the blood. *Spiritus Ammoniæ Fœtidus.* (See *Assafœtida.*) *Ammoniæ Benzoas.* (See *Benzoin.*) *Ammonii Bromidum.* (See *Bromum.*)

LITHIUM. L. 7.

Only two salts of this metal are officinal ·

Lithiæ Carbonas.—Carbonate of Lithia. L_2CO_3.
Characters.—A white powder, or minute crystalline grains; alkaline. Solubility, 1 in 100 of water.
Impurities.—Lime, alumina; detected by lime-water. Deficiency of lithia; detected by weight of residue.
Dose.—3 to 6 gr., in 3 or 4 oz. of aërated water.

Preparation.

a. Liquor Lithiæ Effervescens.—Effervescing Solution of Lithia. Lithia Water. Made like Potash Water. 10 gr. to 1 pint. *Dose,* 5 to 10 fl.oz.
From Lithiæ Carbonas is made :

b. Lithiæ Citras.—Citrate of Lithia. $L_3C_6H_5O_7$.
Source.—Made by dissolving Carbonate of Lithia in a solution of Citric Acid, and evaporating.
Characters.—A white amorphous deliquescent powder. Solubility, 1 in $2\frac{1}{2}$ of water. *Dose,* 5 to 10 gr.

ACTION AND USES.

1. IMMEDIATE LOCAL ACTION.

Externally, lithia may be used as a fomentation in gout.
Internally, lithium has doubtless an **antacid** action on the

alimentary canal very similar to that of potash, but it is not used for this purpose directly.

2. ACTION ON THE BLOOD, AND ITS USES.

Lithium enters the blood, and behaves there much like potash, **increasing its alkalinity**, and combining with such acid bodies as uric acid, for which it has a powerful affinity, (1 part of a solution of the carbonate of lithia, at 38° C., dissolving four parts of the acid). It is extensively used in gout, to hold this substance in solution, and thus prevent acute attacks by fresh deposit in the tissues.

3. SPECIFIC ACTION AND USES.

In this respect also lithia closely resembles potash, being a cardiac and nervo-muscular depressant, if given in large doses or for a length of time; but the risk of lithia poisoning is too small to be allowed to interfere with the exhibition of the drug in suitable cases.

4. REMOTE LOCAL ACTION AND USES.

Lithium is rapidly excreted by the kidneys, and probably by the mucous membranes. It is a powerful **diuretic** in passing through the renal epithelium; and whilst increasing the volume of water, it **diminishes its acidity**, and holds in solution even an excess of uric acid. It is accordingly used as a valuable remedy in gout, as it hastens the excretion of the products which it dissolves in the blood; and in acid lithiasis or gravel, where it prevents the deposit of salts in the kidney and urinary passages.

Both of the salts of lithia may be used, the only important difference between them being with respect to their solubility, which is very marked.

CALCIUM. Ca. 40. LIME.

Creta or chalk is naturally discussed along with calcium or lime, of which it is the carbonate.

The various preparations of lime may be represented as follows, according to their source.

1. Creta.—Chalk. $CaCO_3$ (impure). Native.

    From Creta are made:

    a. **Creta Præparata.**—Prepared Chalk. $CaCO_3$ (nearly pure).

Source.—Made from Chalk by elutriation and subsequent drying.

Characters.—A white amorphous powder, insoluble in water; incompatible with all acids and sulphates.

Impurities.—Alumina and oxide of iron.

Dose.—10 to 60 gr.

Preparations.

α. <u>Mistura Cretæ.</u>—Chalk Mixture. Prepared Chalk, Gum Acacia, Syrup, and Cinnamon Water. 1 in 32. *Dose*, 1 to 2 fl.oz.

β. <u>Pulvis Cretæ Aromaticus.</u>—Aromatic Powder of Chalk. Prepared Chalk, Cinnamon, Nutmeg, Saffron, Cloves, Cardamoms, and Sugar. 1 in 4. *Dose*, 10 to 60 gr.

From Pulvis Cretæ Aromaticus is made :

i. Pulvis Cretæ <u>Aromaticus cum Opio.</u>—Aromatic Powder of Chalk and Opium. 39 parts of Aromatic Chalk Powder, mixed thoroughly with 1 part of Opium in powder. 1 of opium in 40. *Dose*, 10 to 40 gr.

γ. <u>Hydrargyrum cum Cretâ.</u>—Mercury with Chalk. Mercury, 1; Prepared Chalk, 2. *Dose*, 3 to 8 gr.

b. <u>**Calx.**</u>—Lime. CaO (with some impurities).

Source.—Made by calcining Chalk. $CaCO_3 = CaO + CO_2$.

Characters.—Compact whitish masses.

From Calx is made :

α. <u>**Calcis Hydras.**</u>—Slaked Lime. Hydrate of Lime. CaH_2O_2 (with some impurities).

Source.—Made by slaking Lime with distilled water. $CaO + H_2O = CaH_2O_2$.

Characters.—A white powder, strongly alkaline. Soluble in 900 parts of water. Incompatible with vegetable and mineral acids, alkaline and metallic salts, and tartar emetic.

Preparations.

. <u>**Liquor Calcis.**</u>—Solution of Lime. Lime Water. Made by shaking up Slaked Lime in Distilled Water, and decanting after the excess has subsided. $\frac{1}{2}$ gr. of lime in 1 fl.oz. *Dose,* $\frac{1}{2}$ to 4 oz.

From Liquor Calcis is prepared :

Linimentum Calcis. — Liniment of Lime. Solution of Lime and Olive Oil, equal parts mixed.

Liquor Calcis is also used in preparing Lotio Hydrargyri Flava, Lotio Hydrargyri Nigra, and Argenti Oxidum.

ii. **Liquor Calcis Saccharatus.**—Saccharated Solution of Lime. Made by digesting Slaked Lime with Sugar and Water, and straining. 7·11 grains of lime in 1 fl. ounce. *Dose,* 15 to 60 min.

From Calcis Hydras are made :

iii. **Calx Chlorata.**—Chlorinated Lime. " Chloride of Lime." $CaCl_2O_2, CaCl_2$.

Source.—Made by passing Chlorine gas over Slaked Lime. $2CaH_2O_2 + 2Cl_2 = CaCl_2O_2$, $CaCl_2 + 2H_2O$. Partially soluble in water, bleaches and disinfects.

Impurities.— Deficiency in chlorine; detected by volumetric test.

Characters.—A dull white powder, with a feeble odour of chlorine.

Preparations.

(1) Liquor Calcis Chloratæ.—Solution of Chlorinated Lime. 1 pound to 1 gallon of Water; mixed, agitated, and strained. See *Chlorum.*

(2) Vapor Chlori. Inhalation of Chlorine. Chlorinated Lime, moistened with cold water, to disengage chlorine. See *Chlorum.*

Calx chlorata is also employed in preparation of Chloroform. See *Chloroformum.*

iv. **Calcis Hypophosphis.**—Hypophosphite of Lime. $Ca2PH_2O_2$.

Source.—Made by heating Slaked Lime and Water with Phosphorus, purifying the liquid, and crystallising. $3CaH_2O_2 + P_8 + 6H_2O = 3(Ca2PH_2O_2) + 2PH_3$.

Characters. — White pearly crystals, with

a nauseous taste. Soluble in 6 parts of cold water; insoluble in spirit.

Dose.—5 to 10 gr.

Calcis Hypophosphis is used to make :

(1) Sodæ Hypophosphis.—See *Soda.*

c. Calcii Chloridum.—Chloride of Calcium. $CaCl_2$.

Source.—Made by neutralising Hydrochloric Acid with Chalk, and evaporating to dryness. $CaCO_3 + 2HCl = CaCl_2 + CO_2 + H_2O$.

Characters.—White, very deliquescent masses; solubility, 1 in 2 of water.

Impurities.—Carbonates, salts of alumina, and iron; Hypochlorites, detected by evolving Cl with HCl.

Dose.—10 to 20 gr.

Calcii Chloridum is used to make :

a. Calcis Carbonas Præcipitata.—Precipitated carbonate of lime. $CaCO_3$.

Source.—Made by mixing boiling solutions of Chloride of Calcium and Carbonate of Soda, and washing the precipitate. $CaCl_2 + Na_2CO_3 = CaCO_3 + 2NaCl$.

Characters.—A white crystalline powder, insoluble in water. *Dose,* 10 to 60 gr.

Calcis Carbonas Precipitata is contained in Trochisci Bismuthi (4 gr. in each).

2. Calcis Phosphas.—Phosphate of lime. $Ca_3 2PO_4$.

Source.—Made by (1) dissolving Bone-ash in Hydrochloric acid; and (2) adding Solution of Ammonia, and washing and drying the precipitate. (1) $Ca_3 2PO_4 + 4HCl = CaH_4 2PO_4 + 2CaCl_2$; (2) $CaH_4 2PO_4 + 2CaCl_2 + 4NH_4HO = Ca_3 2PO_4 + 4NH_4Cl + 4H_2O$.

Characters.—A light white amorphous powder, insoluble in water.

Dose.—10 to 20 gr.

Calcis Phosphas is contained in :

Pulvis Antimonialis (2 parts in 3). See *Antimonium.*

ACTION AND USES.

1. IMMEDIATE LOCAL ACTION AND USES.

Externally.—Lime in the form of the hydrate is **caustic,** like the alkalies, but its action is more localised, so that it may be combined with fused potash to form a convenient caustic, *potassa cum calce*, Vienna paste, for ordinary use. Dusted on the skin as chalk, or applied in lime-water, it is **astringent and desiccative** (drying), and is used to promote the healing of burns, eczema, and ulcers. The linimentum calcis is a valuable application to burns.

Internally.—The local effect of lime is **antacid**, like the alkalies and magnesia, combined with an astringency peculiar to itself. In the mouth, chalk is used as an antacid and physical dentifrice. Admitted into the stomach and intestines, as lime-water or the carbonate, lime unites with the free acids of the contents, and acts upon the structures of the gastric wall. It is accordingly useful in acid dyspepsia, with heartburn, given after food, *e.g.* as the bismuth lozenge. Lime-water prevents the gastric juice from curdling milk in large lumps, and is extensively given to artificially-reared infants, the liquor calcis saccharatus being an excellent form when dilution of the food is injurious. Its power of combination with acids also makes lime a valuable **antidote** for poisoning by the mineral acids, oxalic acid, and chloride of zinc, and one which is always available in the form of wall-plaster; it must be freely given. The action of lime on the glands of the stomach appears to be depressant, and it is, therefore, not suited for administration before meals. Lime-water is, indeed, a general gastric sedative of some value, arresting some kinds of vomiting, especially in the acid dyspepsia of infants, and in pregnancy.

But little of the alkaline effect of lime or chalk can remain in the bowels beyond what has been exerted on the chyme. The **astringent** action of these familiar drugs in diarrhœa may be in part due to their antacid property, in part to an obscure sedative effect on the intestinal glands (? and vessels), which diminishes the excretion of water into the bowel. The lime salts can be traced along the whole length of the canal, and most of their bulk is finally expelled unabsorbed. Lime and chalk thus come to be two of our most valuable astringents in diarrhœa, either alone or with aromatics, opium, or vegetable astringents, as in the officinal preparations.

Lime-water is also employed locally as an enema for killing the thread-worm, and as a vaginal injection in leucorrhœa.

2. ACTION ON THE BLOOD.

Lime enters the circulation in very small quantities only, and appears in the serum as the phosphate. It probably increases the alkalinity somewhat, but no special use is made of it for this purpose.

3. SPECIFIC ACTION AND USES.

The important part played by lime as a constituent of bones has suggested its use as a specific remedy in rickets, fractures, and other lesions of these structures; and the phosphates and lime-water are extensively used for the two former conditions. The phosphate and the chloride have been recommended in scrofulous diseases of glands and phthisis, to promote absorption, or possibly to induce calcification; and apparently with some reason. None of the calcareous mineral waters, however, are of much service in this respect.

4. REMOTE LOCAL ACTION AND USES.

The greater part of lime or its salts being expelled by the bowel, little remains to be excreted by the kidney, so that their diuretic and alkaline effects are not marked. Their most useful form for influencing the urine is Bath or Wildungen water.

5. ACTION AND USES OF THE DIFFERENT SALTS OF LIME.

Creta in its various forms and combinations—*Liquor Calcis* and *Liquor Calcis Saccharatus*—possesses the general effects and uses of lime. *Calcii Chloridum* is a gastric sedative, but is also a specific in scrofulous enlargement of glands; and is used in testing. *Calcis Phosphas* is a specific in bone diseases and scrofula. *Calx Chlorata* and its derivates are media for supplying chlorine, and used accordingly. *Calcis Hypophosphis* is employed as a specific in tuberculosis and other wasting diseases. In the remaining preparations the action of the lime or chalk is comparatively insignificant, as, *e.g.*, in the three preparations of mercury, of which they are ingredients, and in antimonial powder.

MAGNESIUM. Mg. 24.

The following are the officinal salts and preparations of magnesium, arranged according to their source:

Magnesiæ Sulphas.— Sulphate of Magnesia. Epsom Salt. $MgSO_4.7H_2O$.

Source.—Native.

Characters.—Minute colourless rhombic prisms, with a bitter taste. Solubility, 10 in 13 parts of cold water. Incompatible with alkaline carbonates, lime-water, acetate of lead, and nitrate of silver.

Impurities.—Lime, iron, and general impurities.
Dose.—60 gr. to ½ oz.

Preparations.

a. Enema Magnesiæ Sulphatis.—1 oz., in 1 oz. of Olive Oil, and 15 oz. of Mucilage of Starch; for one enema.

b. Mistura Sennæ Composita.—1 oz. in 5 fl.oz. See *Senna.*

From Magnesiæ Sulphas are also made :

c. **Magnesiæ Carbonas.**—Carbonate of Magnesia (heavy). $(MgCO_3)_3.MgO.5H_2O$.

Source.—Made by mixing boiling solutions of Sulphate of Magnesia and Carbonate of Soda, evaporating, purifying, and drying. $4MgSO_4 + 4Na_2CO_3 + H_2O - 3MgCO_3,Mg2HO + 4Na_2SO_4 + CO_2$.

Characters.—A white granular powder, insoluble in water.

Impurities.—Lime, sulphates, metals.

Dose.—10 to 60 gr.

Preparations.

a. **Liquor Magnesiæ Carbonatis.**—Solution of Carbonate of Magnesia. "Fluid Magnesia." $MgCO_3 + H_2CO_3$ or MgH_22CO_3, in solution.

Source.—Made by passing an excess of Carbonic Acid gas through a mixture of Carbonate of Magnesia (freshly prepared) and water. 1 fl.oz. contains 13 gr. of carbonate.

Characters.—A clear fluid, sometimes effervescing slightly.

Dose.—1 to 2 fl.oz.

β. Trochisci Bismuthi.—2½ gr. in each.

From Magnesiæ Carbonas is made :

Magnesia.—Magnesia (heavy). Magnesia Ponderosa. MgO.

Source.—Made by heating Carbonate of Magnesia in a crucible to expel the carbonic acid.

Characters.—A white powder, comparatively insoluble in water (1 part in 5,412 cold water, 1 in 36,000 hot water).

Impurities.—Those of the carbonates.

Dose.—10 to 60 gr.

d. **Magnesiæ Carbonas Levis.**—Light Carbonate of Magnesia. $(MgCO_3)_3.MgO.5H_2O$.

Source.—Made like magnesiæ carbonas, but with cold dilute solutions instead of hot.

Characters.—A very light white powder, proving microscopically to be partly amorphous, partly prismatic crystals. Soluble in 2,493 parts of cold water, or in 9,000 parts of hot water; 3½ times the bulk of the heavy carbonate.

Dose.—10 to 60 gr.

From Magnesiæ Carbonas Levis is made :

Magnesiæ Levis.—Light Magnesia. MgO.

Source.—Made by heating Light Carbonate of Magnesia in a crucible to expel the carbonic acid.

Characters.—A bulky white powder, 3¼ times the bulk of heavy magnesia.

Dose.—10 to 60 gr.

Magnesiæ Carbonis Levis is contained in Pulvis Rhei Compositus (6 parts in 9).

ACTION AND USES.

1. IMMEDIATE LOCAL ACTION AND USES.

Externally, magnesia has no action, and is not used.

Internally, magnesia is a valuable means of decomposing the contents of the stomach and intestines under various circum stances. The base and carbonates form comparatively insoluble or innocuous compounds with the mineral acids, oxalic acid, mercuric, arsenical, and cupric salts; in large quantities they prevent the absorption of alkaloids by rendering the contents of the stomach alkaline; whilst the sulphate precipitates insoluble sulphates of lead and baryta. Magnesia or its salts may therefore be employed as antidotes in cases of poisoning by any of these substances, the oxide being preferred to the carbonate, so as to prevent the evolution of carbonic acid, and care being taken to give it very freely.

By a similar process of decomposition, magnesia neutralises normal or excessive acidity in the stomach and bowels, and is itself converted into the chloride, lactate, and bicarbonate, this reaction removing irritant acid, and forming salts of magnesia, which have a stimulant or purgative action on the intestine. The carbonate is similarly decomposed, yielding carbonic acid, which exerts its specific action on the stomach. Both substances are therefore employed as local alkaline remedies in acidity of the stomach (heartburn, pyrosis, etc.), given with sal-volatile, after meals, a further laxative effect on the intestine being intended.

The chloride, bicarbonate, or lactate formed in the stomach, and the sulphate of magnesia directly given, having reached the intestine, are very slowly absorbed, and if in sufficient quantity, produce very marked local effects as **saline purgatives**, the sulphate being hydragogue in its action. The result is the free evacuation of a quantity of water by the bowel, and with it the whole, or almost the whole, of the magnesia. Sulphate of magnesia (Epsom salt) is our most common saline purgative, used in the form of Mistura Sennæ Composita (black draught), of a simple solution in sulphuric acid and water with some carminative, and of several of the popular aperient waters, such as Friedrichshall, Püllna, Hunyàdi Jànos, of all of which it is an important constituent. Sulphate of magnesia is regarded as a mild, painless, non-nauseating purgative, less rapid in its action than the soda salt, to be used for completing the effect of purgative pills, for congestion of the portal system, for chronic constipation as an habitual laxative in combination with other salts in the above-named waters, and for feverish attacks with loaded bowels.

Magnesia and the carbonates, when used as purgatives, are chiefly given to children in diarrhœa with foul acid stools, very frequently in the form of Pulvis Rhei Composita (Gregory's Powder). In small doses, neither salt has any purgative action on the bowel, but enters the blood.

2. ACTION ON THE BLOOD AND ITS USES.

Entering the circulation as the chloride or lactate, magnesia **increases** the natural **alkalinity** of the plasma, of which it is a normal constituent, and helps to hold in solution any acid which may be in excess. It will therefore be useful in gout, lithiasis, and possibly in chronic rheumatism, to assist the more powerful alkalinisers of the blood with which it is combined in the waters of Ems, Baden-Baden, Aix-les-Bains, Carlsbad, etc.

3. SPECIFIC ACTION AND USES.

Magnesia taken medicinally does not exert any appreciable effect upon the tissues or nutrition generally. Although an important constituent of bone, it cannot be said to be of any value in rickets or other diseases in which the osseous tissue is deficient in solid matter.

4. REMOTE LOCAL ACTION.

When magnesia does not purge, it is excreted chiefly by the kidneys, rendering the urine more abundant and less acid, and dissolving uric acid. Its **diuretic** and **alkalinising** effects contribute to the value of magnesia waters in gout and gravel.

BARIUM. Ba. 137.

Barium is chiefly used in testing.

Chloride of Barium, $BaCl_2.2H_2O$.
Characters.—Colourless translucent tables.
Dose.—$\frac{1}{2}$ to 2 gr.

Solution of Chloride of Barium, 1 in 10.

ACTION AND USES.

Baryta greatly disturbs the blood pressure, and the chloride has accordingly been recommended in aneurism.

GROUP II.
THE METALS.

ALUMINIUM. Al. 27·5.

Only one salt of this metal is used in medicine.

Alumen.—Alum. $NH_4Al(SO_4)_2.12H_2O$. A sulphate of ammonia and alumina, crystallised from solution in water.
Characters.—Colourless transparent octahedra, with an acid, sweetish, astringent taste. Solubility, 1 in 12 of cold, and in $\frac{2}{3}$th parts of boiling water. *Incompatible* with alkalies, lime, barytas, lead, tartrates, and tannic acid, mercury, and iron.
Impurity.—Sulphate of Iron
Dose.—10 to 20 gr.

From *Alumen* is made :

Alumen Exsiccatum.—Dried Alum.
Source.—Made by heating alum up to 400° till aqueous vapours cease to be disengaged, and powdering.
Characters.—An opaque white bulky powder, or spongy masses. Has lost 47 per cent. by heating; difficult of solution in water, but unites readily with it.

ACTION AND USES.
1. IMMEDIATE LOCAL ACTION AND USES.

Externally.—Alum is astringent, an effect which is fully discussed under *Plumbum.* The dried alum absorbs water, and is somewhat caustic if the skin be broken, for instance, over ulcers.   It is used to destroy weak exuberant granulations.

Internally.—The local action of alum is appreciated in the mouth as an " astringent taste," and in the throat as " dryness," the mucous secretions of the parts being coagulated, and the membrane constringed, especially if the parts be inflamed and swollen.   Alum is therefore a remedy for sore throat, in the form of gargles or sprays, variously combined with other substances.

A similar effect is produced on the mucous membrane of the stomach and intestines, dyspepsia and constipation being the result ; in large doses it is an emetic, irritant, and purgative.   A teaspoonful mixed with syrup is an excellent vomit in croup.   As an injection in discharges from the vagina, uterus, and urethra, it is in constant use ; and also as a wash for the eyes.

2. ACTION IN THE BLOOD.

Alum is absorbed into the blood, probably as an albuminate.

3. SPECIFIC ACTION AND USES.

This salt is believed to possess astringent properties when it reaches the tissues, arresting hæmorrhage and chronic inflammatory discharges from the mucous membranes; and is used in the treatment of hæmoptysis, epistaxis, gleet, diarrhœa, and even whooping cough.   Much of this is doubtful.

4. REMOTE LOCAL ACTION AND USES.

Alum is excreted by the kidneys, and may arrest hæmorrhage from these organs.   Part of the salt may also escape by the skin, as it proves useful in some cases of excessive sweating.

PLUMBUM.   Pb. 207.   LEAD.

Lead is one of the most powerful and useful of metallic drugs.

1. Plumbi Oxidum.—Oxide of Lead. Litharge. PbO.
Characters.—Heavy scales of a pale brick-red colour.   Soluble in diluted nitric and acetic acids.

Impurities.—Copper, iron, and carbonates; detected by ordinary tests.

Preparations.

Emplastrum Plumbi. — Lead Plaster.   1 in $2\frac{1}{2}$ of Olive Oil and 1 of Water.

Plumbi Oxidum or its Emplastrum is also contained in Emplastra Ferri, Galbani, Hydrargyri, Resinæ, Cerati Saponis, and Saponis.

From Plumbi Oxidum is made :

Plumbi Acetas.—Acetate of Lead. "Sugar of Lead." $Pb(C_2H_3O_2)_2.3H_2O$.

Source.—Made by dissolving Oxide of Lead in Acetic Acid and Water, and crystallising. $PbO + 2C_2H_4O_2 = Pb(C_2H_3O_2)_2 + H_2O$.

Characters.—White spongy-looking masses of interlaced acicular crystals, slightly efflorescent, having an acetous odour and a sweet astringent taste. Solubility, 10 in 25 of water.

Incompatibles.—Hard water, mineral acids and salts, vegetable acids, alkalies, lime-water, iodide of potassium, all vegetable astringents, preparations of opium, albuminous liquids.

Impurity.—Carbonate; detected by turbidity of aqueous solution.

Dose.—1 to 4 gr.

Preparations.

a. Pilula Plumbi cum Opio.—Acetate of Lead, 6 ; Opium, 1 ; Confection of Roses, 1. A 4-gr. pill contains 3 gr. of plumbi acetas, and $\frac{1}{2}$ gr. of opium.

Dose.—3 to 5 gr.

b. Suppositoria Plumbi Composita.—Each 15-gr. suppository contains 3 gr. of acetate of lead, and 1 gr. of opium.

c. Unguentum Plumbi Acetatis.—12 gr. to 1 oz. benzoated lard.

d. Liquor Plumbi Subacetatis.—"Goulard extract." $Pb_2C_4H_6O_5$, dissolved in water.

Source.—Made by boiling acetate of lead and oxide of lead in water.

Characters.—A dense, clear, colourless liquid, with a sweet astringent taste and alkaline reaction.

From Liquor Plumbi Subacetatis are prepared :

α. Liquor Plumbi Subacetatis Dilutus.—"Goulard Water." Solution of Subacetate of lead, 1 ; Rectified spirit, 1 ; Water, 78.

β. Unguentum Plumbi Subacetatis Compositum.—1 in $5\frac{3}{4}$.

2. Plumbi Carbonas.—Carbonate of Lead. "White Lead." $PbCO_3$.

Characters.—A heavy white powder. Insoluble in water.
Impurities.—Lime ; sulphate of lead.

Preparation.

Unguentum Plumbi Carbonatis.—1 in 8.

3. Plumbi Nitras.—Nitrate of Lead. $Pb(NO_3)_2$.

Source.—Made by dissolving lead in boiling nitric acid slightly diluted, and crystallising out.
Characters.—Colourless octahedral nearly opaque crystals.

From Plumbi Nitras is made :

Plumbi Iodidum.—Iodide of Lead. PbI_2.
Source.—Made by mixing solutions of Nitrate of Lead and Iodide of Potassium. $Pb2(NO_3) + 2KI = PbI_2 + 2KNO_3$.
Characters.—A bright yellow powder or crystalline scales.

Preparations.

a. Emplastrum Plumbi Iodidi.—1 in 9.
b. Unguentum Plumbi Iodidi.—1 in 8.

ACTION AND USES.

1. IMMEDIATE LOCAL ACTION AND USES.

Externally.—Lead is said not to be absorbed by the unbroken skin ; yet the dilute solution of the subacetate (Goulard water) is of unquestionable value in the treatment of contusions and superficial inflammations, such as erysipelas, and is extensively used in these conditions. In the same form, or as the ointment of the subacetate, it relieves itching ; a symptom, the cause and pathological nature of which are still obscure. Lead may be absorbed when applied in the form of ointment, probably by decomposition ; and the specific effects to be presently described may arise in this way. Lead salts act readily upon wounds, ulcers, and exposed mucous membranes : (1) precipitating the albuminous fluids which cover their surface, or are flowing from them as a discharge ; (2) coagulating the protoplasm of the young cells of the superficial layers ; and (3) contracting the small arteries and veins of the part, thus diminishing or even arresting the circulation within them, and preventing the escape of the plasma and blood-cells through their walls ; whilst (4) the nerves are probably also depressed. These effects are called, as a whole, **astringent**, **antiphlogistic**, and **sedative**. The solutions of the subacetate are much employed as applications to ulcers, as injections for chronic inflammatory discharges from the vagina, urethra, ulcers, ear, etc., and as a collyrium for the conjunctiva; or the carbonate may be dusted upon ulcers, or used as ointment. The strong solution of the subacetate is a powerful irritant, causing pain and reaction, and is rarely used undiluted. The nitrate is stimulant or even caustic, and is applied to syphilitic onychia and chapped nipples. The iodide, in the form of the unguentum, may be rubbed into enlarged joints, glandular swellings and nodes, its absorptive effect being chiefly referable to the iodine.

Internally.—The local action of lead is first appreciated in the mouth as a peculiar " astringent taste," with a sharp sweetness in the case of the acetate. On the mucous membrane of the throat it acts as already described, coagulating the mucus, producing an astringent effect on the cells and vessels of the part, and causing a sensation of dryness. If inflammation be present it is rapidly controlled; and the subacetate, either painted on in the form of the strong solution, or as a gargle formed of the weak solution, is an efficacious remedy for tonsillitis.

The local action of lead on the stomach and intestine corresponds with what has been described: it diminishes the secretions, contracts the vessels, and arrests or retards the peristaltic movements; whilst it is itself converted into an albuminate by the fluids which it encounters. The acetate is accordingly given with or without opium to arrest hæmatemesis; and it is one of the most certain drugs in the treatment of obstinate diarrhœa, especially if ulceration be present, and hæmorrhage threatening, as in typhoid fever (where it may be advantageously combined with opium), and in tuberculosis of the bowels.

2. ACTION IN THE BLOOD.

Lead enters the blood as albuminate, but passes very rapidly through it, and cannot be found in it even after large doses. If lead be given for some time, the blood becomes more watery, and the red corpuscles fewer in number.

3. SPECIFIC ACTION.

All the tissues take up lead freely from the blood, and retain it obstinately as albuminate, the central nervous system being the important seat of its deposit, whilst it is even more abundant in the kidneys and liver as the channels of its escape, and in the bones from the sluggishness of their metabolism. Thus combined with the active cells of the body, lead after a time sets up a series of phenomena known as "plumbism." These are pathological, not physiological, effects, and may be briefly said to take the form of dyspepsia, constipation, and colic; a full, tense, and infrequent pulse, with increased cardiac action; disturbances of the urinary flow; neuralgic pains; tremors, followed by paralysis of the muscles, chiefly affecting the extensors of the wrist; anæmia and emaciation.

These symptoms and the results obtained by experiments on animals have been variously interpreted. Some authorities refer them to an irritant action of lead on the involuntary muscular fibres of the stomach, bowels, blood-vessels, similar to its astringent local effects, whence muscular contractions, painful spasms, narrowing of the vessels, and finally paralysis, and other phenomena from exhaustion. Other pharmacologists contend that lead acts primarily on the central nervous system and nerves, and secondarily only on the muscles, vessels, etc. Its remarkable effect in raising the blood pressure has been referred to irritation of the splanchnics, and consequent narrowing of the abdominal vessels; that is, to increased peripheral resistance. The increased blood pressure is the cause of the infrequent powerful cardiac action, and to some extent of the urinary disturbances.

4. SPECIFIC USES.

The specific action of lead is turned to many important uses. As a powerful **hæmostatic** it is used in bleeding from the stomach and bowel, as we have said, and also from the lungs, opium being advantageously combined with it to ensure mental and bodily rest (Pilula Plumbi cum Opio, or acetate of lead and acetate of morphia with acetic acid). Its use in diarrhœa is also partly referable to its specific action.

5. REMOTE LOCAL ACTION AND USES.

Lead is slowly excreted in the bile, urine, skin, and milk. In the bowel, the portion that has been excreted by the liver is reabsorbed, is again excreted, and finally escapes in the fæces as the black sulphide. In passing through the kidneys, lead diminishes the excretion of uric acid. It is used as a hæmostatic in renal hæmorrhage, in bronchorrhœa, and in profuse sweating.

6. ACTION AND USES OF THE DIFFERENT SALTS OF LEAD.

The special action and uses of the different preparations of lead are as follows: The *Acetate* is the only salt given internally. The solutions of the *Subacetate* are the only liquid preparations of the metal, and are used externally as lotions, injections, collyria, etc., as well as in the form of the ointment. The *Oxide* is made into *Emplastrum* plumbi, the basis of almost all plasters. The *Nitrate* is used as a local stimulant or escharotic, as described; and pharmaceutically to obtain the *Iodide.* The latter possesses, as already described, absorptive powers by virtue of the iodine, an effect which the lead probably promotes, *Plumbi Carbonas* is applied, either as the powder or as an ointment, for astringent purposes, to ulcers and inflamed surfaces.

ARGENTUM. 108. SILVER.

Two salts of this metal are officinal, the nitrate and the oxide.

Argentum Purificatum.—Refined Silver. Pure Metallic Silver.
Impurities.—Lead and copper.

From Argentum is made :

Argenti Nitras.—Nitrate of Silver. $AgNO_3$. Lunar Caustic.
Source.—Made by dissolving Silver in Diluted Nitric Acid.
Characters.—Colourless tabular right rhombic prisms, or white cylindrical rods. Solubility, 100 gr. in 50 min. of water.
Impurities.—Other nitrates; detected by evaporation of filtrate after precipitation with HCl.
Dose.—$\frac{1}{6}$ to $\frac{1}{3}$ gr.

From Argenti Nitras is made :

Argenti Oxidum.—Oxide of Silver. Ag_2O.
Source.—Made by precipitating a solution of Nitrate of Silver with Lime-Water—$2AgNO_3 + Ca2HO = Ag_2O + Ca2NO_3 + H_2O$.
Characters.—An olive-brown powder; insoluble in water. *Incompatible* with creasote, with which it forms an explosive substance.
Impurities.—Metallic silver, evolving gas with nitric acid.
Dose.—$\frac{1}{2}$ to 2 gr.

ACTION AND USES.

1. IMMEDIATE LOCAL ACTION AND USES.

Externally.—In the form of the solid pencil, nitrate of silver is a caustic causing destruction, with deep staining of the superficial layers, acute pain, inflammation of the deeper layers, separation of the part as a slough, and then rapid healing. Unlike potash, its effects are limited to the area of application. On this account it is the best caustic for ordinary use, to destroy the affected part in bites of dogs, serpents, and other venomous animals, in post-mortem wounds and chancres, or to remove small growths. Solutions of the nitrate, when applied to the broken skin or a mucous membrane, exert much the same action as lead, but in a greater degree: precipitating the albumins and the chlorides of the plasma or secretions; coagulating the protoplasm of the young cells of the part; causing active contraction of the arteries, veins, and capillaries; and very rapidly coagulating the blood both within and without them. Nitrate of silver is therefore the best local **antiphlogistic** known, controlling the exudation, growth, and vascular disturbance of the inflammatory process. It is employed to touch callous and weak ulcers, including bed-sores; to control local inflammations in accessible parts; and, as an injection, to wash inflamed surfaces, for example, the urethra, vagina, os uteri, bladder, and conjunctiva. A weak solution is used to harden the skin in threatening bed-sores. Solid caustic is an excellent **hæmostatic** on bleeding from leech-bites.

Internally.—In the mouth, silver causes a nauseous astringent metallic taste. Meeting with chlorides and albuminous fluids, it combines with these, and acts upon the surface of the mucous membrane as it does upon the skin. It is a useful remedy in inflammation of the tonsils and pharynx, whether applied in the solid form as an antiphlogistic in acute cases, or in solution as an **astringent** in relaxed, chronic states.

Reaching the stomach, nitrate of silver is decomposed by the hydrochloric acid and mucus, and cannot act as an irritant upon the mucous membrane unless given in poisonous doses. Its use in ulcers of the stomach must therefore be questioned. When given for ulceration of the bowels, it is administered *per rectum*.

2. ACTION IN THE BLOOD.

Silver enters the blood either as albuminate, or is absorbed as the pure metal by the intestinal epithelium and lacteals, after the manner of fat.   It has no appreciable effect on the blood.

3. SPECIFIC ACTION AND USES.

Silver becomes locked up in all the connective tissues of the body, in the metallic form, staining exposed parts a dusky black brown, incapable of removal.   It probably, therefore, remains inert within the body; but some authorities believe that it affects the nervous tissues, and recommend it in epilepsy, chorea, and locomotor ataxy.   The permanent unsightly discoloration of the skin, which comes on after its use for several months, is a serious objection to its employment.

4. REMOTE LOCAL ACTION.

As we have just seen, silver once admitted to the tissues is not excreted.   A certain amount has, however, been found in the urine; and a proportion always passes through the bowels unabsorbed, appearing on the fæces as sulphide.   No use is made of these facts.

5. ACTION AND USES OF THE DIFFERENT SALTS OF SILVER.

The *Nitrate* is almost invariably used both externally and internally.   The *Oxide* is less irritant, and is chiefly given internally in the form of pill.

CUPRUM.   63·5.   COPPER.

The sulphate is the only salt of copper employed medicinally, although other compounds, as well as the metal itself, are introduced into the Pharmacopœia as tests.

1. Cupri Sulphas.—Sulphate of Copper. $CuSO_4.5H_2O$.

Source.—Made by heating Sulphuric Acid with Copper, dissolving the soluble product, evaporating and crystallising.

Characters.—A blue crystalline salt in oblique prisms. Solubility, 1 in 3 of water.

Impurity.—Iron.

Dose.—As an astringent or tonic, $\frac{1}{4}$ to 2 gr.; as an emetic, 5 to 10 gr.

From Cupri Sulphas is prepared:

Sulphate of Copper, Anhydrous. $CuSO_4$.—A yellowish-white powder made by heating sulphate of copper to 400° Fahr.   Used in testing.

F—8

Copper wire is used for preparing Spiritus Ætheris Nitrosi.

2. Subacetate of Copper of Commerce. Verdigris. Œrugo. $(C_2H_3O_2)_2Cu,CuO$. In powder, or masses of very minute crystals of a bluish-green colour. Used in chemical testing.

ACTION AND USES.

1. IMMEDIATE LOCAL ACTION.

Externally.—The action of copper differs but little from that of silver and zinc. It does not affect the unbroken skin, nor is it absorbed by it into the blood. Applied freely to wounds, ulcers, or the delicate surface of exposed mucous membranes, such as the conjunctiva, the sulphate or "blue-stone" is **caustic**, and is in frequent requisition to destroy warts, chancres and poisoned wounds, and for similar purposes. A swift and slight application of the crystal, or its solution in water, acts so far like nitrate of silver—precipitating the discharges from a mucous or ulcerated surface; coagulating the superficial layers ; thus contracting the blood-vessels and arresting discharge. It is used as a **stimulant** to ulcers; and a solution of 2 to 5 gr. to the oz. may be used as an **astringent** lotion, or injected into the vagina, rectum, or urethra.

Internally.—The local action of copper on the mouth, beyond its astringent metallic taste, corresponds with that just described. If long administered, it may cause a greenish discoloration of the bases of the teeth (*not* of the gums) from direct combination with decomposing products there.

Sulphate of copper, in large doses (10 gr.), is not entirely converted into an albuminate in the stomach, but acts on the mucous membrane as an irritant, and causes vomiting. It is a rapid **direct emetic**, and is suited for administration when the stomach is to be surely and speedily emptied of a narcotic poison, such as opium, or the air-passages evacuated of mucus or false membrane, as in bronchitis and diphtheria, after ipecacuanha has failed. It causes less depression and subsequent nausea than tartar emetic. If sulphate of copper fail to vomit, it must be evacuated by some other means, otherwise dangerous inflammation may result.

Lastly, copper sulphate is a valuable antidote to phosphorus, as it is reduced by the metalloid, the copper being deposited upon the phosphorus and rendering it inert. In cases of poisoning by phosphorus, 3 gr. of blue-stone should be given in water every few minutes, until vomiting occurs, whereupon a free saline purgative is to be administered.

In the intestines copper is an astringent in small quantities, an irritant purgative in larger quantities. Small doses, combined usually with opium, are given for some kinds of diarrhoea.

2. ACTION IN THE BLOOD.

Given in small doses, copper is absorbed into the blood; but we neither know any effect that it produces here, nor use it in this connection.

3. SPECIFIC ACTION AND USES.

The specific action of copper on the tissues is most difficult to evoke, as anyone cán testify who has watched a large number of persons working in brass and copper. It is said to weaken the voluntary muscles and heart, and to affect the nutrition of the central nervous system; whence it was formerly used in convulsions and spasmodic diseases, including epilepsy, chorea, and hysteria. This treatment is now almost obsolete. It is believed by some to be a specific astringent to the uterus.

4. REMOTE LOCAL ACTION.

Copper is chiefly excreted by the liver, that is, leaves the body with the bile and fæces; part is discharged in the urine, and part by the saliva. No special advantage is taken of its elimination by these channels.

ZINCUM. ZINC. 65.

Zincum Granulatum.—Made by pouring fused zinc into cold water.

From Zincum Granulatum are made:

a. Zinci Chloridum.—Chloride of Zinc. $ZnCl_2$.

Source.—Made by dissolving Zinc in Diluted Hydrochloric Acid, and evaporating; Chlorine Water and Carbonate of Zinc being used in the process, to precipitate iron or tin present as impurities.

Characters.—Colourless rods or tablets, very deliquescent, and caustic. Solubility, 10 in 4 of water, freely in rectified spirit and ether.

Impurities.—Sulphates, iron, and calcium.

b. Liquor Zinci Chloridi.

Source.—Made as above, without evaporation.

Characters.—Colourless. Used externally only.

c. Zinci Sulphas.—Sulphate of Zinc. $ZnSO_4,7H_2O$.

Source.—Made from Zinc and Sulphuric Acid, like the Chloride, with the same precautions.

Characters.—Colourless prisms, with a metallic styptic taste.

Impurities.—Iron, lead, copper, arsenic.

Dose.—1 to 3 gr. as a tonic; 10 to 30 gr. as an emetic.

From Zinci Sulphas are made :

a. **Zinci Carbonas.** — Carbonate of Zinc. $ZnCO_3(ZnO)_23H_2O.$ "Calamine."

Source.—Made by decomposing a solution of Sulphate of Zinc with a solution of Carbonate of Soda. (1) $ZnSO_4 + Na_2CO_3 = ZnCO_3 + Na_2SO_4.$ (2) $3ZnCO_3 + 3H_2O = ZnCO_32(ZnO),3H_2O + 2CO_2.$

Characters. — A white, tasteless, inodorous powder, insoluble in water; an impure carbonate.

Impurities.—Sulphates, chlorides, copper.

From Zinci Carbonas are made :

(i) **Zinci Oxidum.**—Oxide of Zinc. $ZnO.$

Source.—Made by heating the Carbonate.

Characters.—A soft, nearly white, tasteless, and inodorous powder, insoluble in water.

Impurities. —The carbonate ; effervescing with acids. And its impurities.

Dose.—2 to 10 gr.

Preparation.

Unguentum Zinci Oxidi.—80 gr. to 1 oz. Benzoated Lard.

(ii) **Zinci Acetas.** — Acetate of Zinc. $Zn(C_2H_3O_2)_2.2H_2O.$

Source.—Made by dissolving Carbonate of Zinc in Acetic Acid and Water, and crystallising. $ZnCO_3,2ZnO.3H_2O + 6C_2H_4O_2 = 3(Zn2C_2H_3O_2) + 6H_2O + CO_2.$

Characters. — Thin colourless crystalline plates, of a pearly lustre, with sharp, unpleasant taste.   Solubility, 10 in 25 of water.

Impurities.—Those of the carbonate.

Dose.—1 to 2 gr. as tonic ; 10 to 20 gr. as an emetic.

Zinci Carbonas is also used in making Zinei Chloridum and Zinci Sulphas.

β. **Zinci Valerianas.**—Valerianate of Zinc. $Zn(C_5H_9O_2)_2.$

Source.—Made by mixing solutions of Sulphate of Zinc and Valerianate of Soda, and crystallising.

Characters. — Brilliant white, pearly, tabular crystals, with an odour of valerianic acid, and a metallic taste. Solubility, 1 in 120 of water; 1 in 60 of spirit.

Impurities.—Sulphate and butyrate of zinc.

Dose.—1 to 3 gr.

Non-officinal Preparations of Zinc.

Calamina Præparata.—Calamine. Impure Oxide of Zinc, obtained by calcining native Carbonate of Zinc, and reducing it to an impalpable powder. A greyish or flesh-coloured powder.

Oleate of Zinc.—Made by heating Oxide of Zinc with Oleic Acid. 1 to 8.

Incompatibles of Zinc Salts in general.

Alkalies and their carbonates, lime-water, acetate of lead, nitrate of silver, astringent vegetable infusions or decoctions, and milk.

ACTION AND USES.

1. IMMEDIATE LOCAL ACTION AND USES.

Externally.—The salts of zinc closely resemble in their action the salts of lead, silver, and copper, being caustic in their stronger forms, astringent or antiphlogistic in their weaker forms. Zinc presents every degree of this action, according to the salt employed, that is probably according to the solubility and diffusion-power of the particular combination of the metal. Thus the chloride, which is highly deliquescent, penetrates the tissues, and is a powerful escharotic, causing destruction of the part, with severe pain, separation of a slough, and subsequent healing. It is employed to destroy morbid growths, chronic ulcers, and gangrenous parts, in the form of a paste or of solid arrows made with plaster of Paris or flour, or as a strong solution. The sulphate and acetate have less affinity for water, and are much less powerful than the chloride. When applied to the broken skin, an ulcer, or an exposed mucous surface, they precipitate the albuminous juices or secretions, coagulate the protoplasm of the upper layers of growing cells, and indirectly cause contraction of the vessels, though less than silver and lead. The sulphate of zinc is the most common of all applications for healing ulcers and wounds, limiting the amount of discharge, checking excessive or " weak " growth, and modifying the intensity of the inflammatory process with which the healing is associated. A solution of this salt is the basis of the ordinary " red lotion " of many hospital pharmacopoeias; and other weak solutions of the same may be employed as a wash or injection for the eyes, urethra, vagina, and other accessible mucous tracts. The oxide and carbonate of zinc, and calamine, act locally as mild astringents in inflamed conditions of the superficial layers of the skin, such as eczema, controlling exudation and hyperæmia, and protecting the parts from the air. Being insoluble in water, they are applied in the form either of powder or of the ointment.

Internally, the local action of zinc corresponds. It is but little used in the mouth or throat, but its effect on the stomach as a local irritant furnishes us with the most familiar of our direct emetics. Sulphate of zinc, in doses of 20 grains, causes rapid and complete vomiting, attended by less immediate depression and less subsequent nausea than antimony and ipecacuan. It is much employed in narcotic poisoning; more rarely in croup, diphtheria, and phthisis, to clear the air passages; or even to empty the stomach in painful dyspepsia. The oxide on reaching the stomach is dissolved, and acts like the soluble salts of zinc.

In the intestine the irritant action of zinc is continued, if it be given in large doses, but this effect is never desired therapeutically. On the contrary, the oxide, in sufficient doses to relieve a moderate superficial catarrh, is often a very efficacious astringent in the treatment of diarrhœa in children.

2. ACTION IN THE BLOOD.

Zinc readily enters the circulation, but nothing is known respecting its influence on the plasma or corpuscles which can be turned to therapeutical account.

3. SPECIFIC ACTION AND USES.

The action of zinc upon the tissues has been learned chiefly from its effect in certain diseased conditions in man, and is but imperfectly understood. It appears to be a depressant to the nervous and muscular systems, and has been employed with unquestionable success in epilepsy, chorea, and whooping cough, all of which are characterised by nervo-muscular excitement. Observations on animals, in which the irritability of the voluntary and cardiac muscles is found to be decidedly reduced by zinc, confirm this experience.

4. REMOTE LOCAL ACTION AND USES.

The kidneys and mammary gland, and probably the mucous surfaces and skin, are the channels of elimination of zinc. It is possible that the metal exerts a second or remote **astringent** effect on these parts as it is leaving the system; for the sulphate and oxide appear to have the power of arresting chronic discharges from remote mucous passages, such as the uterus and vagina, even when given internally; and it is certain that the oxide diminishes the perspirations of phthisis in some instances.

5. ACTIONS AND USES OF THE DIFFERENT SALTS OF ZINC.

These have been sufficiently indicated in the preceding description. The *Chloride* stands alone as a powerful escharotic, never to be given internally; it possesses also disinfectant properties, as the *Liquor Zinci Chloridi*, which is used to mop out very foul wounds, and very extensively to wash infected rooms, flush drains, etc. The *Sulphate* and *Acetate* closely resemble each other in their action, but the acetate is little used. The *Oxide* and *Carbonate* are similarly allied to each other, the former being generally employed. *Zinci Valerianas* probably acts as a zinc salt only, the valerianic acid appearing to be inert. See *Valerianæ Radix.*

CADMIUM. CADMIUM. Cd. 112.

Cadmii Iodidum.—Iodide of Cadmium. CdI_2.
Source.—Made by direct combination of Iodine and Cadmium in the presence of Water.
Characters.—Flat white micaceous crystals, of a pearly lustre. Solubility, 1 in $1\frac{1}{2}$ of water.
Impurity.—Zinc, and general impurities.

Preparation.
Unguentum Cadmii Iodidi.—1 in 8.

ACTION AND USES.

Cadmium closely resembles zinc in its action, both locally and specifically, but is even more **irritant**, and is not given internally. The iodide, in the form of the ointment or in aqueous solution, is applied as a **local stimulant** to enlarged joints and glands, to promote absorption, instead of the iodide of lead which stains the skin yellow.

CERIUM. Ce. 92.

Only one salt of this metal is officinal.

Cerii Oxalas.—Oxalate of Cerium. $CeC_2O_4 3H_2O$.

Source.—Made by precipitating a solution of Oxalate of Ammonia with a soluble salt of Cerium.

Characters.—A white granular powder; insoluble in water.

Impurities.—Alumina; detected by its solution in potash giving precipitate with NH_4Cl. Other oxalates, the ash of which gives effervescence with boiling HCl.

Dose.—1 to 2 gr.

ACTION AND USES.

Nothing is definitely known about the physiological action of cerium. It is given with benefit in vomiting, acid dyspepsia, and heartburn, especially when they occur in pregnancy; and has been credited with good effects in chronic nervous diseases such as epilepsy and chorea.

⌄ FERRUM. Iron. Fe = 56.

Wrought-iron in the form of wire or nails free from oxide. Iron wire No. 35.

Preparations.

1. Mistura Ferri Aromatica.—Iron Wire, 2, Pale Cinchona Bark, 4; Calumba, 2; Cloves, 1; Compound Tincture of Cardamoms, 12; Tincture of Orange Peel, 2; and Peppermint Water, 50.

Dose.—1 to 2 fl.oz.

2. Vinum Ferri.—Iron wire digested in Sherry, 1 in 20

Dose.—1 to 4 fl.dr.

From Ferrum are also made:

3. Ferri Sulphas.—Sulphate of Iron. $FeSO_4.7H_2O$.

Source.—Made by dissolving Iron Wire in Sulphuric Acid and Water, and crystallising.

Characters.—Pale green rhombic prisms, with a styptic taste. Solubility, 1 in $1\frac{1}{2}$ of water; insoluble in spirit.

Impurities.—Persalts, giving sediment in aqueous solution. Copper, precipitated by H_2S.

Dose.—1 to 5 gr.

Preparations.

From Ferri Sulphas are made:

a. **Ferri Sulphas Exsiccata.** $FeSo_4.H_2O$.—A dirty white powder, made by heating the Sulphate.

Dose.—$\frac{1}{2}$ to 3 gr.

b. **Ferri Carbonas Saccharata.**—37 per cent. of Carbonate of Iron, $FeCO_3$, mixed with Peroxide of Iron and Sugar.

Source.—Made by precipitating a solution of Sulphate of Iron with Carbonate of Ammonia, and rubbing with Sugar. (1) $FeSO_4 + (NH_4)_2CO_3 = FeCO_3 + (NH_4)_2SO_4$. (2) $3FeCO_3 + O = FeCO_3 + Fe_2O_3 + 2CO_2$. The sugar prevents oxydation.

Characters.—Grey-brown lumps, with a sweet chaly-beate taste.

Impurities.—Sulphate of ammonia and oxide of iron.

Dose.—5 to 20 gr.

Preparation.

Pilula Ferri Carbonatis.—4 to 1 of Confectio Rosæ.

Dose.—5 to 10 gr.

c. **Mistura Ferri Composita.**—"Griffiths' Mixture." Sulphate of Iron, 25 gr. ; Carbonate of Potash, 30 gr. ; Myrrh, 60 gr.; Sugar, 60 gr.; Spirit of Nutmeg, 4 fl.dr. ; Rose Water, 9 fl.oz. *Dose*, 1 to 2 fl.oz.

d. **Ferri Arsenias.**—Arseniate of Iron. $Fe_3As_2O_8$.

Source.—Made by precipitating a mixed solution of Arseniate and Acetate of Soda with Sulphate of Iron; and washing. $3FeSO_4 + 2Na_2AsO_4 + 2NaC_2H_3O_2 = Fe_3As_2O_8 + 3Na_2SO_4 + 2C_2H_4O_2$.

Characters.—A green amorphous powder, tasteless (but not to be tasted), insoluble in water.

Impurities.—Sulphates, and general impurities.

Dose.—$\frac{1}{16}$ to $\frac{1}{2}$ gr. in pill.

e. **Ferri Phosphas.**—Phosphate of Iron. $Fe_3P_2O_8$.

Source.—Made by precipitating a mixed solution of Phosphate and Acetate of Soda with Sulphate of Iron. $3FeSO_4 + 2Na_2HPO_4 + 2NaC_2H_3O_2 = Fe_3P_2O_8 + 3Na_2SO_4 + 2C_2H_4O_2$.

Characters.—A slate-blue amorphous powder, in-soluble in water.

Impurity.—Arsenic; detected by Reinsch's test.

Dose.—5 to 10 gr.

f. **Ferri Persulphatis Liquor.**

Source.—Made by boiling nitric acid and water with a hot solution of sulphate of iron in sulphuric acid and water. $6(FeSO_4) + 3(H_2SO_4) + 2HNO_3 = 3(Fe_2 3SO_4)$

$+ 4H_2O + 2NO.$ Introduced only for making several preparations.

Characters.—Dark brown, inodorous and astringent.

From Liquor Ferri Persulphatis are made:

α. Ferri Oxidum Magneticum.—Magnetic Oxide of Iron. Fe_3O_4, with some peroxides.

Sources.—Made by precipitating a solution of the Proto- and Persulphates of Iron with a Solution of Soda, and drying. (1) $FeSO_4 + Fe_23SO_4 + 8NaHO = Fe2HO + Fe_26HO + 4Na_2SO_4.$ (2) $Fe2HO + Fe_26HO = Fe_3O_4.4H_2O.$

Characters.—A brownish-black tasteless powder.

Impurity.—Metallic iron; detected by effervescing with HCl.

Dose.—5 to 10 gr.

β. Ferri Peroxidum Humidum.—Moist Peroxide of Iron. $Fe_2.6HO.$

Source.—Made by precipitating Solution of Persulphate of Iron with Solution of Soda. $Fe_23SO_4 + 6NaHO = Fe_26HO + 3Na_2SO_4.$

Characters.—A soft reddish-brown mass.

Impurities.—Ferrous hydrate. Ferric oxyhydrate; insoluble in cold HCl.

Dose.—$\frac{1}{4}$ to $\frac{1}{2}$ oz.

From Ferri Peroxidum Humidum is made:

Ferri Peroxidum Hydratum.—Hydrated Peroxide of Iron. $Fe_2O_22HO.$ Made by drying the Moist Peroxide. A reddish-brown powder, without taste. *Dose,* 5 to 30 gr.

Impurities.—Ferrous hydrate.

Preparation.

Emplastrum Ferri. — "Emplastrum Roborans." 1 in 11.

From Ferri Peroxidum Hydratum is made:

Ferrum Redactum. — Reduced Iron. Metallic iron, with a variable amount of Magnetic Oxide.

Source.—Made by passing dry Hydrogen over the Hydrated Peroxide.

Characters.—A fine greyish-black powder.

Impurity.—Excess of oxide; detected volumetrically.
Dose.—1 to 5 gr.

Preparation.

Trochisci Ferri Redacti.—1 gr. in each.

γ. Ferri et Ammoniæ Citras.—Citrate of Iron and Ammonia.

Source.—Made by dissolving Hydrated Peroxide of Iron (freshly prepared from the Solution of the Persulphate by Ammonia) in a hot solution of Citric Acid, neutralising with Ammonia, and evaporating.

Characters. — Deep red scales, deliquescent; slightly sweet and astringent in taste. Solubility, 10 in 5 of water, almost insoluble in rectified spirit.

Impurities.—Tartrates; giving crystalline precipitate with acetic acid. Alkaline salts; detected in ash.

Dose.—5 to 10 gr.

Preparation.

Vinum Ferri Citratis.—1 gr. in 1 fl.dr. of orange wine. *Dose*, 1 to 4 fl.dr.

δ. Ferri et Quiniæ Citras.—Citrate of Iron and Quinia.

Source.—Made like Ferri et Ammoniæ Citras, freshly precipitated Quinia being also dissolved in the Citric Acid solution.

Characters.—Greenish-yellow scales, deliquescent, bitter and chalybeate in taste. Solubility, 2 in 1 of water. 6 gr. contain 1 grain of quinia.

Impurities.—Alkaline salts; detected in ash. Other alkaloids instead of quinia; insoluble in ether when precipitated by NH_4HO.

Dose.—5 to 10 gr.

ε. Ferrum Tartaratum.—Tartarated Iron.

Source.—Made like Ferri et Ammoniæ Citras with Acid Tartrate of Potash instead of Citric Acid.

Characters.—Garnet scales. Solubility, 1 in 4 of water; sparingly in spirit.

Impurities.—Ammonia; evolved by boiling with liquor sodæ. Ferrous salts.

Dose.—5 to 10 gr.

ζ. **Tinctura Ferri Acetatis.** — Tincture of Acetate of Iron.

Source.—Made by adding Acetate of Potash and Rectified Spirit to the Solution of the Persulphate, and filtering. $Fe_23SO_4 + 6KC_2H_3O_2 = Fe_26C_2H_3O_2 + 3K_2SO_4$.

Characters.—A deep brown liquid,

Dose.—5 to 30 min.

4. Ferri Sulphas Granulata.—Granulated Sulphate of Iron. $FeSO_4.7H_2O$.

Source.—Made by pouring a solution of Iron Wire in Diluted Sulphuric Acid into Rectified Spirit, stirring the mixture; and drying.

Characters.—Small, pale green granular crystals. Solubility, 1 in $1\frac{1}{2}$ of water, insoluble in spirit.

Impurities.— Same as of Ferri Sulphas.

Dose.—1 to 5 gr.

Preparation.

Syrupus Ferri Phosphatis.—1 gr. of Ferri Phosphas in 1 fl.dr.

Source.—Made by precipitating a solution of Granulated Sulphate of Iron with a mixed solution of Phosphate and Acetate of Soda, dissolving the precipitate in Diluted Phosphoric Acid, and adding Sugar.

Characters.—Colourless, becoming brown. 1 fl.dr. contains 1 gr. of Ferri Phosphas ($Fe_3P_2O_8$).

Dose.—1 to 4 fl.dr.

5. Ferri Iodidum.—Iodide of Iron. FeI_2.

Source.—Made by boiling Iron Wire and Iodine in Water, and evaporating.

Characters.—Crystalline, green with a tinge of brown, inodorous, deliquescent. Consists of iron, 1; iodine, $4\frac{1}{2}$; water, $1\frac{1}{4}$. Solubility, 1 in 1 of water.

Dose.—1 to 5 gr.

6. Pilula Ferri Iodidi.—Made by mixing a solution of Iron Wire and Iodine in Water, with Sugar and Liquorice. 1 in $3\frac{1}{2}$. *Dose*, 3 to 8 gr.

7. Syrupus Ferri Iodidi.—Made by mixing a solution of Iron Wire and Iodine in Water with Syrup. Colourless. 4·3 gr. in 1 fl.dr. *Dose*, 20 to 60 min.

8. Ferri Perchloridi, Liquor Fortior.—Strong Solution of Perchloride of Iron.

Source.—Made by adding Hydrochloric Acid and Nitric Acid to a hot solution of Iron Wire in Hydrochloric Acid and Water (1) $Fe + 2HCl = FeCl_2 + H_2$. (2) $6FeCl_2 + 6HCl + 2HNO_3 = 3Fe_2Cl_6 + 4H_2O + 2NO$.

Characters.—An orange-brown liquid, consisting of perchloride of iron, Fe_2Cl_6, in solution in water. Not given internally.

Impurities.—Ferrous salts.

Preparations.

a. **Liquor Ferri Perchloridi.**—Pale brown. 1 to 3 of water. *Dose,* 1^0 to 30 min.

b. **Tinctura Ferri Perchloridi.**—Light brown. 1 to 3 of spirit. *Dose,* 10 to 30 min.

9. Liquor Ferri Pernitratis.—Solution of Pernitrate of Iron.

Source.—Made by dissolving Iron Wire in Diluted Nitric Acid.

Characters.—A clear reddish-brown liquid, slightly acid and astringent to the taste; consisting of pernitrate of iron, Fe_26NO_3, in solution in water.

Impurities.—Ferrous salts.

Dose.—10 to 40 min.

Incompatibilities of Preparations of Iron in General.

Alkalies and their carbonates, lime-water, carbonate of lime, magnesia, and its carbonate, give green precipitates with proto-salts, brown with persalts. Tannic and gallic acids give a deep blue-black with persalts. Preparations of iron therefore tinge infusion of chiretta and hops, and change to brown or black those of chamomile, cusparia, gentian, orange, cascarilla, cloves, digitalis, cinchona, and all astringent infusions; but they can be given in infusion of quassia or calumba.

ACTION AND USES.

1. IMMEDIATE LOCAL ACTION AND USES.

Externally.—A solution of iron has a corrugating and astringent effect upon the broken skin and mucous surfaces, coagulating the albuminous tissues and plasma, and constringing or condensing the elements. The blood-vessels are thus closed or diminished in size, not actively, but by compression from without; the circulation through them is diminished; hæmorrhage, if present, is arrested; and the abnormal escape of plasma and leucocytes, which characterises chronic inflammation or catarrh, is checked. Solutions of the ferric salts are therefore used as hæmostatics to arrest hæmorrhage from accessible parts, such as leech-bites, the nose, and uterus; less extensively in chronic discharges from the vagina, rectum, and nose, as astringents. Injected into the rectum, they destroy worms. Iron is not absorbed by the unbroken skin.

Internally.—The constringent effect of iron is appreciated in the mouth as a "styptic taste." Beyond this, the local action corresponds with that just described externally. Various iron solutions are usefully applied, either as gargles or with the brush, in some forms of chronic sore throat.

In the stomach all the salts of iron, whatever their nature, are converted into the chloride, and do not combine with the acid albuminates, like some of the other metals. Deficiency of hydrochloric acid or of food, or excess of iron, thus decomposes the gastric juice, and allows the iron to act upon the mucous membrane as an astringent and irritant. Iron is thus directly **unfavourable to** digestion; and in this connection we must carefully note (1) that iron may disorder the digestion even in healthy subjects; (2) that it must not be given for disease until the gastric functions have been so far restored; (3) that it is well to begin then with the mildest preparations; and (4) that it must be given after meals. Iron is a valuable antidote in arsenical poisoning, the humid peroxide ($Fe_2 6HO$) forming with arsenious acid an almost insoluble compound ($As_2O_3 + 2(Fe_2 6HO) = Fe_3As_2O_8 + Fe2HO + 5H_2O$). Abundance of the iron should be given, and the compound should be quickly expelled by a smart purge of sulphate of magnesia or soda. The solutions of the persalts are used to arrest hæmorrhage from the stomach. In the duodenum iron is converted into an alkaline albuminate, and thus absorbed. The further effect of iron on the bowel is a remote one, to be presently described.

2. ACTION ON THE BLOOD, AND ITS USES.

The action of iron on the blood is almost unique of its kind: first, because its specific action is exerted, not upon the plasma, but upon the red corpuscle, and on this alone, not on any other tissue or organ; secondly, because this action appears to be nothing more than the combination of the iron as one of the constituent elements of the corpuscle with the others. In the case of no other metal can we speak so definitely of its *modus operandi.*

Iron enters the circulation along the whole length of the alimentary canal as the chloride and alkaline albuminate, and quickly unites with the corpuscles, as it cannot be found in the plasma. It combines with the hæmoglobin, and as such alone exists in the blood. In normal blood a " course " of iron increases the richness of the blood ; whilst in anæmia the rapidity of the growth of corpuscles and of the rise in value of the hæmoglobin, as estimated day by day with the hæmoglobinometer and hæmacytometer, is remarkable. Iron is accordingly used as a hæmatinic in an endless variety of conditions in which hæmoglobin is deficient, such as simple anæmia, scrofula, amenorrhoea, cardiac disease, syphilis, malarial cachexia, and convalescence from acute disease. The cautions already given respecting digestion must be faithfully respected, to secure its hæmatinic action over a length of time. Iron is an important constituent of many well-known mineral waters, the most important being those of Spa, Tarasp, Kissingen, Kreutznach, Pyrmont, and St. Moritz on the Continent; Harrogate, Moffat, and Strathpeffer in this country; and the Rawley Springs, Sweet Chalybeate, and Bedford, in the United States.

3. SPECIFIC ACTION AND USES.

Iron has no specific action on the organs apart from the blood; and the tonic effect which it produces so satisfactorily, appears to be entirely referable to its action on the blood. Abundance of oxygen is essential for every bodily and mental function; and the feeling of " tone," vigour, and mental fitness varies with the degree of oxygenation of the blood, *i.e.* with the quality of the blood as regards hæmoglobin. Nervous, muscular, and cardiac debility are thus removed by iron, and even digestion is restored by this gastric irritant, if it can be successfully introduced into the blood. The temperature is said to be slightly raised by iron, showing increased oxydation. Iron has also a specific effect in erysipelas, diphtheria, and other adynamic diseases, which cannot be perfectly explained. Fever is generally held to contraindicate the use of iron; and the same may be said of phthisis, except as mild forms in chronic cases.

4. REMOTE LOCAL ACTION AND USES.

Iron is excreted by almost every possible channel. As it is absorbed, so a portion of it is excreted, along the whole length of the intestine, and colours the fæces black (sulphide). Only a small amount escapes in the urine, saliva, sweat, the milk in women, the pancreatic juice, and by the various mucous surfaces. Whilst passing out of the system, iron produces a second or remote effect of an astringent kind. As regards the bowels, the clinical applications of this action are most important. Thus most of the salts of iron cause constipation unless combined with a purgative, such as the sulphates of magnesia and soda, or aloes; no good can be derived from iron until the bowels have been thoroughly relieved, and are acting regularly; and certain salts, such as the perchloride and pernitrate, which are more astringent to the intestines than others, may sometimes be employed to check chronic diarrhoea and dysentery, and to arrest hæmorrhage from the bowel in typhoid fever. The remote astringent action of iron is increased from the fact that it is also excreted by the liver, and passes down with the bile. The urine falls somewhat in volume, but the urea and other solids, as well as the acidity, are increased. Hæmorrhage from the kidney or bladder is arrested by iron, which is also beneficial in some cases of Bright's disease.

Iron similarly reduces the secretion of milk in nursing women. The remote effect of iron on the mucous surfaces renders it a valuable hæmostatic in recurrent passive bleedings from the nose, uterus, and respiratory passages. As a remote astringent, it is invaluable in chronic discharges from the same and allied parts, cspecially in leucorrhoea.

5. ACTIONS AND USES OF THE DIFFERENT PREPARATIONS OF IRON.

Large as is the number of the preparations of iron, they and their special actions may be easily remembered if classified as follows :

1. **Iron, its Oxides and Carbonates.**—This group comprises Ferrum Redactum, Mistura Ferri Aromatica, Vinum Ferri, Ferri Carbonas Saccharata, Mistura Ferri Composita, Ferri Peroxidum Hydratum, and Ferri Peroxidum Magneticum. These preparations possess the hæmatinic action of iron with but little astringency, and are accordingly selected to restore the blood, when the patient has a tendency to dyspepsia and constipation. They are the principal forms of iron used in the routine treatment of anæmia, amenorrhoea, and chlorosis in young women. Let it be observed that these solid preparations form the soluble compounds in the stomach, for absorption into the blood, as readily as do the fluid preparations belonging to the second class. The Mistura Ferri Composita, although a preparation of the protosulphate, contains the carbonate and peroxide, and is a favourite and valuable preparation for anæmia with amenorrhoea; the iron acting as a hæmatinic, the potash also building up the red corpuscle (the salts of which are almost entirely potassium compounds), and the myrrh possibly increasing the production of leucocytes for conversion into the red, as well as stimulating the uterus. Ferrum Redactum, the Saccharated Carbonate and the Hydrated and Magnetic Oxides, although bulky powders, are easily taken. Vinum Ferri is an agreeable

preparation largely prescribed for children. The Aromatic Mixture, containing cinchona and aromatic bitters, is a valuable stomachic tonic and hæmatinic.

2. **Compounds of Iron with the Mineral Acids.**—Ferri Sulphas in its various forms, Liquor Ferri Perchloridi and its preparations, and Liquor Ferri Pernitratis, are comprised in this group, which are characterised by their corrugating and astringent action. They are, therefore, chosen in all the external and internal applications of iron for local purposes, especially as hæmostatics. The strong solution of the perchloride is injected into the uterus in *post partum* hæmorrhage in the form of a watery solution (1 part to 3) with the best results. Cotton wool or lint soaked in the same solution is used for plugging deep wounds, the cavities of the nose, mouth, etc., in hæmorrhage; but the action of the iron on the surfaces of wounds, and the extensive coagulation which it sets up in the veins, are both objections to its employment, unless the bleeding cannot otherwise be arrested. Internally these astringent preparations may be given in hæmorrhage from the stomach or bowels, kidneys or bladder; but not, as a rule, in hæmoptysis. As hæmatinics, the tincture or liquor of the perchloride and the pernitrate, well diluted, are much given to convalescents after the appetite has been restored, and to persons who require a tonic, as well as in passive hæmorrhages and chronic inflammatory discharges, such as leucorrhœa. In ordering this class of iron salts, we must carefully observe the various precautions already mentioned in connection with digestion. Protosulphate is well borne in the form of pill (Blaud).

3. **Compounds of Iron with Vegetable Acids.**—These are the Ferri et Ammoniæ Citras, Ferrum Tartaratum, and Tinctura Ferri Acetatis. They are at once the weakest, the blandest, and the least constipating preparations of iron, and are therefore employed when only small quantities of the metal have to be given over a length of time as a tonic, or to commence a course of hæmatinics when the alimentary canal cannot tolerate the stronger preparations. They make but little impression upon the more severe forms of anæmia. They can be given with alkalies.

4. **Compounds of Iron with other active bodies.**—Iron is combined in the Pharmacopœia with iodine—Ferri Iodidum; with arsenic acid—Ferri Arsenias; with phosphoric acid—Ferri Phosphas; and with quinine—Ferri et Quiniæ Citras. Speaking generally, it may be said that in these preparations the iron is intended to relieve anæmia, or to act as a tonic in the sense we have described, whilst the other constituent is specifically influencing the diseased condition on which the anæmia or

debility depends. Thus the iodide of iron is employed in syphilis and scrofula; the arseniate in chronic diseases of the skin, liver, etc., with a gouty, rheumatic, or malarial taint; the phosphate in diseases of the bones, such as rickets; and the compound with quinine in malarial cachexia, where it may rapidly restore the blood corpuscles. But all the preparations of this group, and especially the last, are also used as ordinary tonics, according to circumstances.

The Solution of the Persulphate of Iron is introduced solely as a source of several other preparations. Ferri Peroxidum Humidum is the best form as an antidote to arsenic, the *rationale* of which has been already explained.

MANGANESIUM. Mn. MANGANESE.

The only salt of this metal in the Pharmacopœia is the black oxide; but permanganate of potash, which is derived from it, is best discussed under this head.

Manganesii Oxidum Nigrum.—Black Oxide of Manganese. MnO_2. A heavy black powder.

From Manganesii Oxidum Nigrum is made:

Potassæ Permanganas.—Permanganate of Potash. $KMnO_4$.

Source.—Made by (1) evaporating a mixture of Black Oxide of Manganese, Chlorate of Potash, and Caustic Potash in water, pulverising the residue, heating it to redness, cooling and pulverising; then (2) boiling in Water, neutralising with Diluted Sulphuric Acid, and evaporating. (1) $3MnO_2 + KClO_3 + 6KHO = 3K_2MnO_4 + KCl + 3H_2O$; a manganate being formed. (2) $3K_2MnO_4 + 2H_2O = 2KMnO_4 + 4KHO + MnO_2$; the manganate becoming permanganate by boiling.

Characters.—Dark purple, slender prisms, inodorous, with a sweet astringent taste, yielding an intense purple solution when moistened. Solubility, 1 in 16 of water. Is very rapidly deoxydised in the presence of organic matter into hydrated peroxide of manganese, losing its purple colour for a brown.

Impurities.—Sulphate of potash, and black oxide of manganese; detected by being less soluble in water, and by volumetric test.

Dose.—1 to 2 gr.

Preparation.

Liquor Potassæ Permanganatis.—4 gr. to 1 fl.oz. *Dose,* 2 to 4 fl.dr.

Manganesii Oxidum Nigrum is also used in making Liquor Chlori and Hydrargyri Perchloridum.

ACTION AND USES.

1. IMMEDIATE LOCAL ACTION AND USES.

Externally.—Permanganate of potash is an irritant or even escharotic in the pure state, stimulant in the form of the solution, and has a healing effect upon scars and wounds. The principal applications, however, are independent of its physiological action on the human tissues, and due to its influence as an **antiseptic, disinfectant,** and **deodorant,** that is, to its action on the processes and products of sepsis, fermentation, and decomposition. By its power of giving up oxygen freely, the permanganate either destroys the ferment or organism on which these processes depend, or forms chemical compounds with the materials on which they flourish (the tissues, plasma, pus, etc.), incapable of decomposition : it is thus an *antiseptic.* By similarly oxydising the products of decomposition already begun, it so alters their chemical properties as to deodorise and decolorise them, and possibly destroys also the power of further infection which such products generally possess : it is thus a *disinfectant.* Permanganate of potash may therefore be used as a dressing for foul ulcers ; but other substances, possessing special advantages, are generally preferred for this purpose.

Internally.—This salt is employed as a mouth-wash in foul condition of the teeth and mouth, as a gargle in putrid sore-throat, and as an injection in infective and foul discharges, such as gonorrhœa, vaginitis, ozœna, and cancer of the uterus.

2. ACTION IN THE BLOOD, SPECIFIC ACTION, AND REMOTE LOCAL ACTION.

Nothing is definitely known of the action of permanganic acid on the blood, tissues, or organs of excretion. It is difficult to believe that any portion of the salt escapes decomposition before absorption, unless given in poisonous doses ; and the oxide of manganese, into which it is converted, is believed to be inert. The internal administration of the potash salt for some supposed effect on infective fevers or gangrenous processes must therefore be useless. It has recently been used as an emmenagogue in amenorrhœa.

By far the most important application of permanganate of potash is as a disinfectant and deodorant, apart from the human body : to disinfect stools and foul discharges after removal from the patient; to wash utensils; and to flush water-closets, etc. Its great advantages are, that it is rapid and complete in its action; odourless and non-poisonous in solutions of ordinary strength; and that it shows by change of colour whether it is acting or exhausted. The principal disadvantage connected with it is its expense.

ᴠHYDRARGYRUM. Hg. 200. Mᴇʀᴄᴜʀʏ.

This metal is of the first therapeutical importance, and a large number of salts and other preparations are made from it.

Hydrargyrum.—Mercury. Hg.
Characters.—A fluid metal, brilliantly lustrous.
Impurities.—Lead, tin, etc.; detected by being non-volatile.

Preparations.

a. **Hydrargryum cum Cretâ.**—1, with 2 of Chalk. *Dose*, 3 to 8 gr.

b. **Emplastrum Hydrargyri.**—1 in 3.

c. **Emplastrum Ammoniaci cum Hydrargyro.** 1 in 5.

d. **Pilula Hydrargyri.**—"Blue Pill." 1 in 3, with Confection of Roses and Liquorice. *Dose*, 3 to 8 gr.

e. **Unguentum Hydrargyri.**—"Blue Ointment." Nearly 1 in 2.

From Unguentum Hydrargyri are prepared :

a. Linimentum Hydrargyri.—1 of Ointment, to 1 of Liquor Ammoniæ, and 1 of Linimentum Camphoræ.

β. Unguentum Hydrargyri Compositum.— 6, with 7½ Olive Oil, Wax, and Camphor.

γ. Suppositoria Hydrargyri.—5 gr. of ointment in each.

From Hydrargyrum are made :

f. **Hydrargyri Oxidum Rubrum.**—Red Oxide of Mercury. HgO.
Source.—Made by triturating together and heating

Mercury, and Mercuric Pernitrate obtained by dissolving Mercury in Nitric Acid. (1) $3Hg + 8HNO_3 = 3(Hg2NO_3) + 2NO + 4H_2O.$ (2) $Hg2NO_3 + Hg = 2HgO + 2NO_2.$

Characters.—An orange-red powder, insoluble in water.

Impurities.—Red lead and brick-dust; detected by being non-volatile. Nitrate of mercury; by yielding nitrous vapours by heat.

Dose.—$\frac{1}{4}$ to 1 gr.

Preparation.

Unguéntum Hydrargyri Oxidi Rubri.—"Red Precipitate Ointment." 1 in 8.

g. **Hydrargyri Iodidum Viride.**—Green Iodide of Mercury. HgI.

Source.—Made by rubbing together Mercury and Iodine in the presence of Rectified Spirit.

Characters.—A dull green powder, insoluble in water.

Impurity.—Biniodide of mercury, found by long keeping; detected by being soluble in ether.

Dose.—1 to 3 gr.

h. **Hydrargyri Sulphas.**—Sulphate of Mercury. $HgSO_4.$

Source.—Made by dissolving Mercury in hot Sulphuric Acid, and drying.

Characters.—A white, heavy, crystalline powder. Used only to prepare calomel and corrosive sublimate.

From Hydrargyri Sulphas are made :

a. **Hydrargyri Subchloridum.**—Subchloride of Mercury. Calomel. HgCl.

Source.—Made by subliming a mixture of Sulphate of Mercury, Mercury, and Chloride of Sodium; and washing with boiling water. (1) $HgSO_4 + Hg = Hg_2SO_4.$ (2) $Hg_2SO_4 + 2NaCl = 2HgCl + Na_2SO_4.$

Characters.—A dull white, heavy, nearly tasteless powder, insoluble in water and spirit.

Incompatible with iodide of potassium, nitro-hydrochloric acid, hydrocyanic acid, solutions of lime, potash, and soda.

Impurities.—Perchloride of mercury; detected by being soluble in warm ether. Other chlorides; non-volatile.

Dose.—$\frac{1}{2}$ to 5 gr.

Preparations.

i. Lotio Hydrargyri Nigra.—Black Wash. Calomel, 30 gr.; Lime Water, 10 fl.oz.

ii. Pilula Hydrargyri Subchloridi Composita.—Plummer's Pill. Calomel, 1; Sulphurated Antimony, 1; Guaiacum Resin, 2; Castor Oil, 1. *Dose*, 5 to 10 gr.

iii. Unguentum Hydrargyri Subchloridi. 1 in $6\frac{1}{2}$.

β. Hydrargyri Perchloridum.—Perchloride of Mercury. "Corrosive sublimate," $HgCl_2$.

Source.—Made by subliming a mixture of Sulphate of Mercury, Chloride of Sodium, and Black Oxide of Manganese. $HgSO_4 + 2NaCl = HgCl_2 + Na_2SO_4$. The manganese simply prevents the formation of calomel.

Characters.—Heavy colourless masses of prismatic crystals. Solubility, 1 in 20 of water.

Incompatible with alkalies and their carbonates, lime-water, tartar emetic, nitrate of silver, acetate of lead, albumen, iodide of potassium, soaps, decoction of bark.

Impurities.—Fixed salts; detected by not volatilising.

Dose.—$\frac{1}{16}$ to $\frac{1}{8}$ gr.

Preparations.

i. Liquor Hydrargyri Perchloridi.—$\frac{1}{2}$ gr. in 1 fl.oz. ($\frac{1}{16}$ gr. in 1 fl.dr.). *Dose*, 30 to 120 min.

ii. Lotio Hydrargyri Flava.—"Yellow Wash." Corrosive Sublimate, 18 gr.; Lime Water, 10 fl.oz.

From Hydrargyri Perchloridum are made :

iii. **Hydrargyri Iodidum Rubrum.**—Red Iodide of Mercury. HgI_2.

Source.—Made by mixing hot solutions of Perchloride of Mercury, and Iodide of Potassium, and purifying the precipitate. $HgCl_2 + 2KI = HgI_2 + 2KCl$.

Characters.—A vermilion crystalline powder. Soluble feebly in water, freely in ether.

Impurities.—As in the perchloride.

Dose.—$\frac{1}{16}$ to $\frac{1}{4}$ gr.

Preparation.

<u>Unguentum Hydrargyri Iodidi Rubri.</u>—1 in 28.

iv. Hydrargyrum Ammoniatum.—Ammoniated Mercury. "White Precipitate." NH_2HgCl.

Source.—Made by precipitating a solution of Perchloride of Mercury with Solution of Ammonia. $HgCl_2 + 2NH_4HO = NH_2HgCl + NH_4Cl + 2H_2O$.

Characters.—An opaque white powder, insoluble in water, spirit, and ether.

Impurities.—As in the perchloride.

Preparation.

<u>Unguentum Hydrargyri Ammoniati.</u>—1 in 8.

v. Hydrargyri Oxidum Flavum.—Yellow Oxide of Mercury. HgO.

Source.—Made by precipitating a solution of Perchloride of Mercury with a Solution of Soda.

Characters.—A yellow powder.

Impurities.—As in the perchloride.

***i.* Liquor Hydrargyri Nitratis Acidus.**—Nitrate of Mercury, $Hg2NO_3$, in solution in nitric acid.

Source.—Made by dissolving Mercury in Nitric Acid and Water.

Characters.—A colourless, strongly-acid liquid.

Impurity.—Subnitrate of mercury; detected by giving precipitate when dropped into diluted hydrochloric acid.

***j.* Unguentum Hydrargyri Nitratis.** — Citrine Ointment. Made by adding Lard melted in Olive Oil to a solution of Mercury in Nitric Acid.

Non-officinal Preparations of Mercury.

<u>Oleate of Mercury.</u>—Made by dissolving 5 to 20 per cent. of Yellow Oxide of Mercury in Oleic Acid.

<u>Donovan's Solution.</u>—Solution of Hydriodate of Mercury and Arsenic. *Dose,* 10 to 30 min.

ACTION AND USES.

1. IMMEDIATE LOCAL ACTION AND USES.

Externally.—Mercury in the form of the acid solution of the nitrate is a powerful caustic, employed to destroy growths on the skin, such as lupus, but must be used with caution. The perchloride applied in weak solutions is not absorbed, but acts destructively on organisms on or in the skin, such as those of ringworm. Stronger solutions cause inflammation of the skin, and concentrated solutions are caustic; but neither effect is surgically employed. A weak solution (gr. $\frac{1}{4}$ to the oz.) is used as a **disinfectant** and **stimulant** to ulcers, acting like other metallic salts (*see* pages 60 and 64), at the same time being absorbed, and producing the specific effects of the metal. Mercury itself, and most of the other preparations, cause little or no irritation of the skin, unless rubbed into it for some time.

The various *methods of administering mercury locally* must here be noticed.

(1) In the form of the ointment, metallic mercury may be applied by *inunction, i.e.* rubbed into a soft part of the skin. Thus applied, mercury undoubtedly enters the blood; but it has been contended that the metal is not admitted by the skin, but through the lungs, in the form of the vapour arising from the heated body smeared with the ointment, or even in small particles by the mouth. Fortunately, the question is of no practical importance, the fact remaining that the system can be quickly brought under the influence of mercury by inunction. The non-officinal oleate *painted* on the skin quickly conveys the metal into the system.

(2) The subchloride (calomel) may be administered by *fumigation.* The vapour of calomel, rising from a vessel heated by a lamp, is conducted to a part or to the whole of the surface of the body of the patient, and there allowed to settle as a fine deposit of the salt. The effect is increased by simultaneous diaphoresis, induced either by the vapour of water or by such a drug as jaborandi. 20 gr. of calomel may thus be fumigated, during a sitting of twenty minutes. The same doubt exists as to the precise way in which the calomel thus applied enters the blood.

(3) As a *bath* of dilute solutions of the perchloride, say 3 dr. to 30 gallons of water, with 1 dr. of hydrochloric acid.

(4) Mercurials may be dusted on to the raw surface of a blistered portion of the skin, or soft syphilitic growths (condylomata)—the *endermic* method, when it is rapidly absorbed.

(5) Solutions of the perchloride (albuminates or peptonates) may be injected *hypodermically*—a powerful method, but apt to produce sores.

(6) The vapour of mercurials may be *inhaled*, as we have seen; but this method is rarely employed intentionally.

(7) Mercury may be given *per rectum*, as the officinal suppositories.

The action of mercury admitted to a part of the body by any of these channels is usually more than local, the specific effects of the drug, presently to be described, being shortly developed. At the same time, the local effect will be more marked: skin diseases will be healed, condylomata removed, and indurations and chronic inflammatory processes reduced in connection with the bones or joints.

Internally.—The local action of mercury is the same as externally, according to the nature and strength of the preparation employed. Very dilute solutions of the perchloride (4 gr. to 10 fl.oz., with 8 min. of hydrochloric acid) may be used as a gargle or wash for syphilitic ulcers of the tongue and gums. All the salts of mercury act upon the mouth, gums, and salivary glands, causing salivation; but this effect is due to their excretion, not to their immediate influence on the parts, and will be described later.

In the stomach, mercurials combine with the chloride of sodium of the secretions, and, whatever their original form, are converted into a double chloride of sodium and mercury, which further unites with the albuminous juices, to form a complex molecule of mercury, sodium, chlorine, and albumen. This compound, although precipitated at first, is soluble in an excess either of chloride of sodium or of albumen; exists in the stomach, therefore, in solution; and is readily diffusible and easily absorbed. It is not specially irritant in moderate quantities, and none of the salts of mercury given in medicinal doses produce vomiting like zinc and copper; indeed, Dr. Ringer has shown that calomel in $\frac{1}{12}$-gr. doses, or Hydrargyrum cum Cretâ in $\frac{1}{3}$-gr. doses, given every two or three hours, arrests some forms of vomiting in children. In large or concentrated doses, however, mercurials are irritant or corrosive to the stomach, and must be given with caution, after meals.

The irritant effect of mercurials continues in the duodenum, naturally taking the form of purgation. The perchloride is never employed to produce this effect, but divided mercury in the form of the Pilula Hydrargyri and Hydrargyrum cum Cretâ, and Calomel, are common purgatives. The action of mercurials as purgatives is a purely local one, none of the metal being absorbed, but the whole expelled in the fæces. The exact nature of this action is, however, obscure. Probably the intestinal glands are chiefly stimulated to increased secretion, and the mucous membrane irritated to such a degree as to produce a moderate increase of watery exudation from its vessels into the bowel, peristalsis becoming more brisk at the same time. The result is a thorough evacuation of the contents of the small intestine as a large, loose, but not watery, stool, charged with bile, which has been hurried out directly from the duodenum, and not allowed to re-enter the portal circulation by absorption from the lower bowel, as it normally does. Thus mercurials, especially calomel, increase the amount of bile evacuated without increasing the amount secreted; that is, are **indirect cholagogues** by being duodenal purgatives. The manner in which indirect cholagogue action stimulates the liver to further secretion is discussed in Part III. The purgative action of mercurials is greatly assisted by a subsequent saline, such as Seidlitz powder, or the Mistura Sennæ Composita. The class of diseases in which mercurials are selected as purgatives chiefly include cases of congestion of the portal system and liver, especially those referable to secondary indigestion from free living or gout; cases of constipation attended by irritable stomach, or actual ulceration of the stomach or bowels; very rarely cases of habitual constipation, except at long intervals, to enable gentle laxative measures to act more freely; and occasionally diarrhœa, when it is distinctly referable to biliary derangement, or the presence of an irritant in the bowel, as in children.

2. ACTION ON THE BLOOD AND ITS USES.

As we have seen, mercury enters the blood freely through the broken or unbroken skin. From the bowel but a small part of a medicinal dose is absorbed, the rest passing off in the fæces as the sulphide, unless combined with opium, which delays its progress through the intestine. The complex molecule which mercury forms in the stomach and intestines is decomposed on entering the blood by combination with oxygen and albumen, an oxyalbuminate of mercury being the result, and apparently the same compound is formed when the metal enters by other channels.

No *direct* effect on the blood can be attributed to mercury; but impairment of nutrition generally, including digestion, attends its excessive use, and induces impoverishment, both of the plasma and the corpuscles, *indirectly* referable to the drug. The blood under these circumstances is more watery and coagulates less firmly, and nutrition may be further disordered in consequence, with the production of low forms of inflammation and ulceration. But it is to be clearly understood that this is not in any sense a specific effect of mercury, and that the influence of mercury upon inflammatory products and syphilitic growths, to be presently described, is not exerted through the blood, but upon the tissues themselves. The impoverishing effect of this drug upon the blood must be constantly kept in mind, and the quality of the blood sustained by abundance of food, and the strictest attention to digestion.

3. SPECIFIC ACTION.

Mercury quickly leaves the blood and enters the tissues, where it is apt to remain almost indefinitely, being excreted with comparative slowness, especially when the kidneys are diseased. It has been found in every organ of the body, most abundantly in the liver. It is a remarkable fact, however, that no definite anatomical change has ever been demonstrated in the viscera, such as the vessels, liver, or nervous system, even in cases of chronic poisoning by this metal; mercury in this respect again differing from lead, silver, antimony, and arsenic. Whilst, therefore, the specific action of mercury is unquestionable, its *mode* of action is still obscure, and numerous theories have been proposed to account for it, which need not be fully discussed here. The most probable explanation of the effects of mercury upon nutrition may be said to be that in some way or other it interferes with the growth or life of germinal cells, and that it has therefore an **alterative** influence on certain processes, such as inflammation and syphilis, which are characterised by a growth of small young cells. Possibly, it may have a destructive influence on certain ferments and organisms connected with physiological and pathological metabolism, one of these being the organism of syphilis.

Whatever may be the explanation of its action, mercury produces a train of symptoms, when given for a considerable period in moderate doses, known as "hydrargyrism," which chiefly take the form of debility; nervous phenomena, including muscular tremors and paralysis, pains, and mental disturbance; cardiac depression; ulceration of the skin, mouth and mucous membranes; salivation, dyspepsia, and diarrhoea. The temperature is not directly raised, nor the excretions increased, so that there is no positive evidence of increased metabolism as an effect of mercury.

4. SPECIFIC USES.

The uses of mercury as a specific remedy bear no definite relation to these effects, which have been mentioned chiefly that they may be recognised and arrested. The principal application of the drug is as an "alterative" in syphilis, a disease attended by the growth of cells around the small vessels, and the development of these into nodes, gummata, various eruptions, etc. Mercury has a powerful influence in controlling the severity of this disease. Its employment may be commenced with various local applications to the primary sore, and regular internal doses of the solution of the perchloride, calomel, grey powder, or some of the other preparations, until salivation threatens. It is generally (not universally) believed that the secondary stage is rendered less severe, or is even entirely prevented by this means. The drug must be continued during the appearance of secondary symptoms; but, as a rule, it is better omitted in the tertiary stage. The particular preparation employed varies with the experience of the practitioner. Quinine and opium are useful means of support to be combined with mercury in a course of the metal, and we must repeat that, unless the appetite and digestion continue good, its use must be interrupted.

The other use of mercurials as alterative remedies is in internal inflammations, especially inflammation of serous membranes, such as peritonitis, pericarditis, pleurisy, meningitis, and orchitis. This line of treatment, once universal in England, is now almost obsolete, excepting, perhaps, in peritonitis of a subacute or chronic kind, in which, as in most instances where it is used as an antiphlogistic, mercury is combined with opium. Possibly some of the benefit thus attending mercurialisation in inflammation, and which was formerly referred to a "resolvent" action on the fibrin of exudations, is due to its purgative and indirect cholagogue effects.

5. REMOTE LOCAL ACTION AND USES.

Mercury passes out of the system in all the secretions—the saliva, sweat, milk, urine, and bile, probably as an albuminate, and stimulates many of the glands *en route*. It is in this way, as we have seen, a powerful **sialagogue**, causing swelling of the salivary glands and a profuse flow of the secretions of the mouth. This effect is important only because it is to be avoided. The diaphoretic effect of mercury is comparatively insignificant. Whilst it does not increase of itself the volume of urine, it assists to a marked degree such diuretics as digitalis and scilla; but it must not be given in kidney disease, as it acts injuriously on the diseased tubules, and readily produces its debilitating effects when the renal function is impaired. In the fæces mercury leaves the body as the sulphide, being derived, first, from that considerable portion of the dose which is not absorbed ; and, secondly, from the portion excreted by the liver (in the bile), and by the pancreas and intestinal glands. It will thus be seen that but little use is made of the remote local action of mercury.

6. ACTION AND USES OF THE DIFFERENT PREPARATIONS OF
MERCURY.

The preparations of mercury, although so numerous, can be readily remembered, and their special actions understood, when they are classified as follows :

1. *Metallic Mercury* and preparations containing it.

2. The *Perchloride* of *Mercury* and its preparations.

3. The *Subchloride* of *Mercury* and its preparations.

4. The *Oxides, Iodides,* the *Ammoniated Mercury,* and their preparations, a complex group, the action and uses of which closely correspond either with those of the perchloride or with those of the subchloride.

5. *Acid Nitrate of Mercury* and the Ointment corresponding.

1. **Metallic Mercury and its preparations.**—These may be employed in all the classes of cases for which mercurials are adapted. The metal itself is never given internally, except in the finely-divided form in which it exists in Pilula Hydrargyri and Hydrargyrum cum Cretâ. The blue pill is chiefly used as a purgative and indirect cholagogue, but is also given in syphilis, in small doses combined with opium and quinine, and in combination with digitalis and scilla as a diuretic (the famous "Guy's pill)." Hydrargyrum cum Cretâ, or "grey powder," is a favourite purgative for children, and also a convenient preparation for a course of mercury in syphilis. Unguentum Hydrargyri, or "blue ointment," is the usual means of administering the metal by inunction in syphilis. A portion as large as a pea or hazel nut is rubbed daily into a soft part of the skin, such as the inside of the thigh, or smeared on flannel, and applied round the loins, the gums being carefully watched. This is a very sure and tolerably safe, but very dirty method, which is chiefly employed with infants. The non-officinal oleate, painted on, is a great improvement in this respect. Mercurial ointment may also be smeared over inflamed parts, such as the testis, and is used as a parasiticide. The Liniment of Mercury (the ointment in a liquid form) is chiefly employed as an antiphlogistic, being soaked on lint and applied to the affected part, *e.g.* the joints or the abdomen in subacute peritonitis, The same use may be made of the plasters, and of the compound ointment, " Scott's dressing." The suppository may be used in syphilis or to kill ascarides.

2. Perchloride of Mercury.—This is the most powerful of all mercurials. It is one of the most active of antiseptics, being 100 times as strong as carbolic acid, and may be used to disinfect foul ulcers, especially of syphilitic origin, a certain amount of caustic and stimulant effect being secured at the same time. It must be cautiously employed. It is also used to destroy the fungus of ringworm. Internally, as the Liquor (a weak solution), it is given in syphilis only, never as a purgative. In this form, the perchloride is by no means an irritant preparation of mercury, but rather the reverse. Lotio Hydrargyri Flava, " yellow wash," containing the yellow oxide, is applied to syphilitic sores.

3. Subchloride of Mercury.—Calomel resembles metallic mercury in being used externally and internally, as a purgative, alterative, and antisyphilitic remedy. Externally it is applied to syphilitic sores and chronic inflammatory growths as calomel dust, by fumigation, as the unguentum, and as the black wash. Internally calomel is a valuable purgative, with the powerful action as an indirect cholagogue and hepatic stimulant already described. The compound calomel pill (Plummer's pill) is in much repute as a hepatic stimulant and alterative, with little or no directly purgative effect, given every night or every other night for a week at a time, in gout and loaded conditions of the system consequent on free living. Calomel, combined with opium, was the favourite mercurial employed by the last generation of surgeons and physicians in the treatment of inflammation, to which we have already referred. In syphilis the same combination is still employed with success.

4. The Oxides, Iodides, and Ammonio-Chloride of Mercury. —These substances, although forming a convenient group, belong, as regard their action and uses, partly to the second and partly to the third group above. Thus the following closely resemble the perchloride, viz. Hydrargyri Oxidum Flavum, Hydrargyri Oxidum Rubrum, Hydrargyri Iodidum Rubrum, and Hydrargyrum Ammoniatum. The first two are almost exclusively used in syphilis, and externally, chiefly according to the opinion and custom of the practitioner. The " white precipitate" ointment is useful as a parasiticide, and as a stimulant application to chronic inflammatory eruptions of almost any kind in children. Along with the subchloride is to be classed Hydrargyri Iodidum Viride, which is much used in syphilis by some surgeons. Donovan's Solution is valuable in obstinate syphilides. The student will not forget that the Lotio Hydrargyri Flava really contains the yellow oxide, and the Lotio Hydrargyri Nigra, the black oxide, although they are reckoned as preparations of the perchloride and subchloride respectively.

5. Liquor Hydrargyri Nitratis Acidus, and the Ointment of the Nitrate.—These are not used in syphilis; but the former is used as a caustic in lupus and other limited growths and ulcers of the skin ; while the ointment is of value as a stimulant in cases of chronic skin disease, and is applied to the edges of the eyelids in chronic inflammation and ulceration of the hair follicles.

Precautions in the use of mercurials.—Mercury must not be given as an alterative, antiphlogistic, or antisyphilitic remedy in persons with anæmia or debility, unless those are distinctly referable to syphilis, and even then it must be employed with caution. Tuberculosis and kidney disease also contra-indicate the use of mercury ; and certain individuals will occasionally be met with in whom even small doses of calomel or blue pill quickly induce hydrargyrism by a kind of idiosyncrasy. In every instance the patient must be carefully nourished, as we have said. On the contrary, children—even infants—bear mercury very well, although the prolonged administration of the metal to them appears to produce a peculiar change in the permanent teeth when they appear, which is extremely unsightly ("mercurial teeth " of Hutchinson).

GROUP III.

THE METALLOIDS.

✓ARSENICUM. Arsenic. As. 75.

Acidum Arseniosum.—Arsenious Acid. White Arsenic. As_2O_3.

Source.—An anhydrous acid, obtained by roasting Arsenical Ores, and purified by sublimation.

Characters.—A heavy white powder, or stratified opaque masses. Solubility, 1 in 100 of cold water ; 1 in 20 of boiling water. *Incompatibles :* Salts of iron ; magnesia, lime-water, and astringent matters.

Impurities.—Lime salts ; detected by non-volatility.

Dose.—$\frac{1}{60}$ to $\frac{1}{12}$ gr. in solution.

Preparations.

a. Liquor Arsenicalis.—" Fowler's Solution."

Source.—Made by dissolving Arsenious Acid and Carbonate of Potash in Water, and colouring with Compound Tincture of Lavender. 4 gr. in 1 fl.oz.

Characters.—A reddish liquid, alkaline to test-paper, with the odour of lavender.

Dose.—2 to 8 min.

b. **Liquor Arsenici Hydrochloricus.**—Hydrochloric Solution of Arsenic.

Source.—Made by boiling Arsenious Acid with Hydrochloric Acid and Water.   No decomposition occurs. 4 gr. in 1 fl.oz.

Characters.—Colourless, with an acid reaction.

Dose.—2 to 8 min.

From Acidum Arseniosum is made :

c. **Sodæ Arsenias.** — Arseniate of Soda. $Na_2 HAsO_4.7H_2O$.

Source.—Made by fusing Arsenious Acid with Nitrate and Carbonate of Soda, boiling the products in Water, and crystallising.   (1) $As_2O_3 + 2Na_2NO_3 + Na_2CO_3 = Na_4As_2O_7 + N_2O_3 + CO_2$.   (2) $Na_4As_2O_7 + 15H_2O = 2(Na_2HAsO_4.7H_2O)$.

Characters.—Colourless transparent prisms.   Solubility, 1 in 2 of water.   The solution is alkaline.

Dose.—$\frac{1}{16}$ to $\frac{1}{8}$ gr.

Preparation.

Liquor Sodæ Arseniatis.—4 gr. in 1 fl.oz.

Dose.—5 to 10 min.

From Arseniate of Soda is made :

Ferri Arsenias.   See *Ferrum.*

Non-officinal Preparation of Arsenic .

Donovan's Solution.   Solution of Hydriodate of Arsenic and Mercury.   *Dose,* 10 to 30 min.

ACTION AND USES.

1. IMMEDIATE LOCAL ACTION AND USES.

Externally.—Arsenious acid is a powerful irritant and caustic.   It is used occasionally to destroy lupus, epithelioma, and other superficial or limited new growths, in the form of " paste," composed of Arsenious Acid (1), Charcoal (1), Red Sulphuret of Mercury (4), and Water.   In the form of a dilute ointment, it is employed in psoriasis to remove the scaly growth.   Arsenic must be used locally with great care, as it is absorbed by the broken skin, ulcers, and mucous membranes, unless sufficient inflammation be set up to throw it off.

Internally.—The local corrosive action of arsenic may be employed in caries of the teeth to destroy the painful pulp before stopping, a paste composed of 2 parts of arsenious acid, 1 part of sulphate of morphia, and a sufficiency of creasote to make a stiff compound, being placed in the cavity.

Reaching the stomach in medicinal doses, the preparations of arsenic do not combine with the albuminous contents like mercury, but remain unchanged. They thus act upon the mucous membrane, stimulating the nerves and vessels, causing a sense of heat and hunger, and increasing the gastric function. In these small doses arsenic is employed with advantage in some cases of gastric dyspepsia, and a similar effect on the duodenum makes it of some value in lienteric diarrhoea. If the dose be increased, the stimulant action passes readily into irritation of the stomach attended by pain, sickness, and diarrhoea from intestinal excitement. These symptoms are to be remembered only that they may be avoided, or arrested if they should arise.

2. ACTION ON THE BLOOD AND ITS USES.

Arsenic enters the blood and combines with the corpuscles, not with the serum, as an albuminate; if in excess, it reduces the number of the blood cells, as well as their oxygenating power. It has been used with success in some forms of anæmia; but less frequently in idiopathic cases than where the corpuscles and plasma have suffered from failure of nutrition elsewhere (symptomatic anæmia), as in tuberculosis, malaria, gout, and rheumatism. Alone or combined with iron, it has sometimes an excellent effect in restoring the blood in such cases.

3. SPECIFIC ACTION AND USES.

Arsenic enters all the organs and tissues, but is not known to combine with their albuminous constituents; it remains in them for a short time only; and is quickly excreted. During this period, however, it distinctly influences metabolism. It first reaches the liver, and diminishes the amount of glycogen in it, so that it may be occasionally, but by no means often, used with success in diabetes. In the other organs it interferes similarly with metabolism, apparently (like phosphorus) through the oxygenating process. An increased amount of nitrogenous waste appears in the urine; the temperature rises; and the excessive fatty product of the albuminous decomposition remains unexcreted, constituting fatty degeneration. Short of this effect, arsenic produces a wholesome increase of the metabolism, or vital activity of all the organs, and is therefore given as a general tonic, and as a valuable **alterative** in such classes of

H —8

disturbed nutrition as gout and chronic rheumatism. It is possible that arsenic affects the life processes of other living particles in the body besides the tissue elements, namely, the organisms of certain diseases. Thus it is, next to quinine, the most successful medicinal agent in the treatment of chronic malaria, brow-ague, and other varieties of neuralgia due to the same cause, and malarial cachexia; and is also used with advantage in hay-fever. It sometimes also dispels lymphomatous tumours. Beyond a safe amount, arsenic produces a series of nutritive disorders in the tissues, characterised chiefly by debility and nervous disturbances, known as "chronic arsenical poisoning," which need not be detailed here.

Next to nutrition generally, the nervous system appears to be most influenced by arsenic, which is found abundantly in the grey matter of the cord in poisoning by this metal. Here it acts by **diminishing** the sensibility and reflex irritability of the centres, as well as of the motor nerves and muscles. Preparations of arsenic are useful in chorea, various forms of neuralgia, and spasmodic asthma, especially when malaria or anæmia, or both, may happen to be associated with the neurosis. Like phosphorus, arsenic is said to cause increase of the compact tissue of bone at the expense of the medullary tissue, but it is not specially used to produce this effect. In large doses it has a depressing effect on the respiration, circulation, and temperature.

4. REMOTE LOCAL ACTION AND USES.

Arsenic is excreted chiefly in the urine in the form of arsenious acid; also by the liver and skin. It is not known to affect the kidney specially, but is sometimes used in chronic Bright's disease. The liver, as we have seen, is modified in its activity; and part of the value of arsenic in chronic gout, gravel, and skin diseases, may be referable to its action on the greatest metabolic organ in the body. Either thus indirectly, or directly, its effect on the skin is so remarkable, that it is the most valuable of all internal remedies for certain eruptions obviously connected with disordered nutrition, such as psoriasis, chronic eczema, acne, and pemphigus, whilst it aggravates such diseases as erythema multiforme. Donovan's Solution is used in syphilides.

5. METHODS OF ADMINISTERING ARSENIC, AND PRECAUTIONS IN ITS USE.

Arsenical preparations should always be given immediately at the end of meals, unless their gastric effect be desired, which is rarely the case; and they ought not to come in contact with the exposed mucous membrane. For the same reason they must not be given as alteratives if dyspepsia be present. Epigastric fulness, pain, and tenderness, a sense of constriction in the throat, irritation or soreness of the conjunctiva, and especially vomiting, ought to suggest a diminution or suspension of the drug. Children bear arsenic with comparative ease, whilst old subjects are said to bear it badly. A combination of iron with arsenic (for example, Vinum Ferri with Liquor Arsenicalis) is one of the best of hæmatinics and tonics, probably because the iron affords a supply of oxygen sufficient to carry to a complete termination the increased metabolism produced by the arsenic.

PHOSPHORUS. P. 31.

A non-metallic element obtained from bones.

Source.—Prepared from Phosphoric Acid or Superphosphate of Lime (obtained by acting on bone-ash by oil of vitriol), by distillation with Charcoal.

Characters.—A semi-transparent, almost colourless, wax-like solid, when fresh; luminous in the dark, ignites in the air; insoluble in water, soluble in ether, oils, and naphtha, entirely soluble in boiling oil of turpentine and bisulphide of carbon.

Preparations.

a. **Oleum Phosphoratum.** — Phosphorated Oil. Made by dissolving Phosphorus in Almond Oil at 180° Fahr. 1 in 160. *Dose, 5 to 10 min.*

b. **Pilula Phosphori.**—Phosphorus, Balsam of Tolu, and Yellow Wax. Apt to pass through the bowels unchanged. *Dose, 3 to 6 gr.* = $\frac{1}{30}$ to $\frac{1}{15}$ gr. of phosphorus.

Phosphorus is also used in preparing Acidum Phosphoricum Dilutum, and Calcis Hypophosphis. See *Calcium.*

ACTION AND USES.

Phosphorus has a powerful action on the body, and one which has been proved by elaborate investigations on animals to be of the most interesting kind to the physiologist. As a poison phosphorus is also of great importance. Unfortunately, however, it cannot be said to be of much value to the therapeutist, as it has disappointed most attempts to turn it to practical account in the treatment of disease.

1. IMMEDIATE LOCAL ACTION.

Externally and internally phosphorus acts as a powerful local **irritant** and **caustic**, and is never given to produce this effect. For the same reason the drug must not be ordered in the solid form, but carefully mixed with oil or fat.

2. ACTION ON THE BLOOD AND ITS USES.

Phosphorus enters the blood, and may be found in it unchanged. Here it is partly oxydised into phosphorus or phosphoric acid at the expense of the oxygen of the red corpuscles, and is therefore said to have a "reducing" action **on** the (**oxy-**) **hæmoglobin** or "blood." The small dose sufficient to cause death will not reduce any considerable number of the corpuscles, and the specific effects to be presently described cannot therefore be accounted for by interference with the oxygenating function of the blood.

Phosphorus has been employed in leukæmia and lymphadenoma, but on the whole with disappointing results.

3. SPECIFIC ACTION AND USES.

In the tissues phosphorus may be traced as the uncombined element—another proof that its oxydation in the blood is incomplete. Its effect on metabolism, when given in large doses, is most distinct and definite : it increases the nitrogenous products, including urea, tyrosin, and leucin; reduces the glycogen of the liver to *nil;* raises the temperature, diminishes the excretion of carbonic acid, and the volume of oxygen absorbed ; and leads to fatty degeneration of epithelial, glandular, and muscular protoplasm throughout the body. No doubt these alterative effects are essentially associated with each other ; phosphorus, whilst increasing metabolism, so influencing it as to diminish oxydation, and thus to arrest the process at the first stage, where proteids are converted into urea and oil, instead of allowing it to proceed to the second or final stage, where the oil is further oxydised into carbonic acid and water. Hence all the results just enumerated ; whilst the soluble products (urea, etc.) are excreted, the insoluble products (oils or fats) are retained in the tissues, constituting fatty degeneration.

The uses to which phosphorus has been put as a specific remedy do not obviously depend upon these effects upon nutrition. It has been given in nervous disorders, such as neuralgia; in adynamic conditions, such as typhoid fever ; in some kinds of skin diseases, including pemphigus ; and as an aphrodisiac. It is difficult to understand how any of these morbid states can be benefited by a substance which diminishes oxydation; and, indeed, the empirical use of phosphorus has recently been in a great measure abandoned.

In very small doses over a considerable length of time, phosphorus affects the structure of bones, converting the spongy portion into firm, compact substance, without in any way altering its composition chemically. It has therefore been recommended in cases of rickets and ununited fracture; but in rickets, at least, is far inferior to other medicinal measures, if of service in any way.

The *hypophosphites* have recently been much employed in cases of nervous and general debility and chronic lung disease, and act, according to some authorities, in the same manner as free phosphorus, without being irritant. As the hypophosphites are probably converted into phosphates in the stomach, they may be expected to stimulate the liver and bowels, and to affect the growth and healing of bones, lymphatic glands, and adenoid tissue, including tubercle.

4. REMOTE LOCAL ACTION.

Phosphorus is excreted by the kidneys as phosphorus and phosphorous acid, not as phosphates; but is not employed in this connection.

ANTIMONIUM. Sb. 122. ANTIMONY.

The metal itself (Stibium) is not officinal, all the preparations being derived from " black antimony," Antimonium Nigrum, as follows :

Antimonium Nigrum.—Black Antimony. Native Sulphide of Antimony. Sb_2S_3.

Source.—Purified from siliceous matter by fusion, and powdered.

Characters.—A metallic-looking powder, of a steel-grey colour.

Impurity.—Silica ; insoluble in boiling HCl.

Not given medicinally.

From Antimonium Nigrum are made :

a. **Antimonium Sulphuratum.**—Sulphurated Antimony. Sulphide of Antimony, Sb_2S_3, with a small and variable amount of Oxide of Antimony, Sb_2O_3.

Source.—Made by (1) boiling Black Antimony with Solution of Soda, and (2) precipitating with Diluted Sulphuric Acid. (1) $2Sb_2S_3 + 6NaHO = 2Na_3SbS_3 + Sb_2O_3$

$+ 3H_2O.$ $(2)\ 2Na_3SbS_3 + 3H_2SO_4 = Sb_2S_3 + 3Na_2SO_4 + 3H_2S.$

Characters.—An orange-red powder, without odour, and with a slight taste, insoluble in water.

Impurities.—General; detected volumetrically.

Dose.—1 to 5 gr.

Antimonium Sulphuratum is an important ingredient of Pilula Hydrargyri Subchloridi Composita. 1 in 5. (See *Mercury.*)

b. **Liquor Antimonii Chloridi.**—Solution of Chloride of Antimony, $SbCl_3$, in Hydrochloric Acid. "Butter of Antimony."

Source.—Made by dissolving Black Antimony in Hydrochloric Acid. $Sb_2S_3 + 6HCl = 2SbCl_3 + 3H_2S.$

Characters.—A heavy yellowish-red liquid, giving a white precipitate with water.

From Liquor Antimonii Chloridi is made :

Antimonii Oxidum.—Oxide of Antimony. $Sb_2O_3.$

Source.—Made by (1) precipitating Oxychloride of Antimony, by pouring the Solution of the Chloride into Water; washing; and (2) adding Carbonate of Soda Solution. (1) $12SbCl_3 + 15H_2O = 2SbCl_3 5Sb_2O_3 + 30HCl.$ (2) $2SbCl_3 5Sb_2O_3 + 3Na_2CO_3 = 6Sb_2O_3 + 6NaCl + 3CO_2.$

Characters.—A greyish-white powder, insoluble in water.

Impurities. — Higher oxides, insoluble when boiled with acid tartrate of potash.

Dose.—1 to 4 gr.

Preparation.

Pulvis Antimonialis—"James's Powder." 1, with 2 of Phosphate of Lime. *Dose,* 3 to 10 gr.

From Antimonii Oxidum is made :

Antimonium Tartaratum. — Tartarated Antimony. Tartar Emetic. $KSbC_4H_4O_7.H_2O.$

Source.—Made by boiling Oxide of Antimony with Acid Tartrate of Potash, and crystallising. $2KHC_4H_4O_6 + Sb_2O_3 = 2KSbC_4H_4O_7 + H_2O.$

Characters.—Colourless transparent crystals, exhibiting triangular facets. Solubility, 1 in 20 of cold water ; 1 in 2 of boiling water.

Incompatibles.—Gallic and tannic acids, most astringent infusions, alkalies, lead salts.

Impurities.—Cream of tartar, and iron; detected volumetrically, and by solubility.

Dose.—As a diaphoretic, $\frac{1}{16}$ to $\frac{1}{8}$ gr.; as an emetic, 1 to 2 gr.

Preparations.

(1) <u>Unguentum Antimonii Tartarati.</u> — to 4.

(2) <u>Vinum Antimoniale.</u>—2 gr. in 1 fl.oz. *Dose*, 5 min. to 1 fl.dr.

ACTION AND USES.

1. IMMEDIATE LOCAL ACTION AND USES.

Externally.—Antimony, in the form of the Liquor Antimonii Chloridi, is a escharotic, employed chiefly in veterinary practice, occasionally by the surgeon as an application to poisoned, foul, or malignant surfaces. Tartarated Antimony applied to the skin, either in aqueous solution or as the officinal ointment (half a drachm at a time, repeated), causes a pustular eruption, and was once used as a counter-irritant in diseases of the lungs, joints, or meninges. Antimony is freely absorbed from the broken skin, and from mucous surfaces.

Internally, the local effect is equally irritant. In doses of 1 to 3 grains tartarated antimony is an emetic, whence its popular name. The effect is partly direct—due, that is, to the irritant action of the drug upon the walls of the stomach; partly indirect, from immediate stimulation of the vomiting centre in the medulla. Further, its direct effect on the stomach is produced not only when the salt is admitted to it by the mouth, but when it reaches the stomach by the blood, that is, when it is being excreted by the gastric mucosa. Thus, whilst tartar emetic induces vomiting most quickly when swallowed, it is not speedy and evanescent in its effects, but induces both previous and subsequent nausea and depression. It is not suited, therefore, for use in cases of poisoning, where rapid evacuation is of the first importance, or where there is much general depression; but is useful in the first stage of acute inflammatory diseases, with sthenic fever, in strong healthy subjects. It is especially indicated in respiratory affections, such as laryngitis and bronchitis, where its remote effects as an expectorant are valuable; or to clear the air-passages in the same diseases or in whooping cough.

In smaller continued doses the local action of tartarated antimony on the stomach and bowels is apt to produce loss of appetite, nausea, pain, and diarrhœa.

2. ACTION IN THE BLOOD.

Antimony enters the blood either from within or from without, but does not appear to combine with the albumen of the plasma. No special action or use has to be mentioned under this head.

3. SPECIFIC ACTION AND USES.

Having reached the tissues and organs, antimony clings to them with some tenacity, and may be found in them months after its administration. Here it sets up a series of important changes, attended by phenomena referable to the general nutrition of the body, the circulation, respiration, and nervous and muscular systems; besides the effects to be afterwards described as referable to its excretion.

The effect of antimony on *metabolism* closely resembles that of phosphorus and arsenic, to the account of which the student is referred. Briefly the principal results are fatty degeneration of the organs and increase of the nitrogenous products, oxygenation being comparatively deficient. Upon this **alterative** effect depends in part the value of antimony in gout, chronic skin disease, etc., to be afterwards described. The *heart* is depressed from the first by tartarated antimony. Even in small doses it reduces the strength, and very soon the frequency of the pulse, which tends to become irregular, and fainting may occur; the whole being referable to a direct action upon the nervo-muscular substance of the heart. Antimony is thus a powerful **circulatory depressant**. The *respiratory* movements are also weakened and disturbed by this drug, which causes shortness of inspiration and lengthening of expiration, manifestly a degree of the same disturbance which culminates in vomiting, and allied to the process of expectoration. The *nervous system* is markedly depressed by antimony, in part directly, in part indirectly through the circulation, the effect of a moderate dose being to produce a sense of languor, inaptitude for mental exertion, lowness, and sleepiness. Tartarated antimony has accordingly been used as a **sedative** in the delirium and insomnia of fevers, such as typhus, and acute alcoholism (delirium tremens), combined with opium in various proportions.

The *muscular system* is so powerfully depressed by antimony that, before the introduction of chloroform, it was employed to produce muscular relaxation in the reduction of herniæ and dislocations. Nauseating and emetic doses cause great weakness of the voluntary movements, inability to stand, occasional tremors, and aching of the muscles. It is still given as an **antispasmodic**, to relax the cervix uteri in some classes of difficult labour, and in combination with purgative medicines to prevent or remove spasm of the bowel.

4. REMOTE LOCAL ACTION AND USES.

Antimony leaves the system by all the mucous surfaces, the liver, kidneys, and skin; so that it may cause inflammation, salivation, and pustulation of the mouth, œsophagus, and stomach when administered by the skin. In being excreted by the *stomach*, it produces there, as we have seen, a remote emetic effect. Its excretion in the *bile* constitutes it a hepatic stimulant, sulphurated antimony, either as Plummer's pill or alone, being much esteemed as a **cholagogue**, especially in gout and loaded conditions of the liver. In passing through the *kidneys*, it has a slight diuretic action. In doses of $\frac{1}{10}$ to $\frac{1}{2}$ gr., it stimulates the skin, acting as a diaphoretic, of service, as we shall see, in feverish conditions. Its internal use occasionally develops the characteristic pustular eruption, which suggests it as a remedy for certain forms of chronic skin disease. Antimonial wine is a familiar **sedative expectorant**, apparently from the excretion of the drug by the respiratory surfaces, given with great advantage in the first stage of acute bronchitis in strong subjects, less frequently in acute pneumonia.

5. USES OF THE COMBINED ACTIONS OF ANTIMONY.

When the various effects of antimony thus detailed are reviewed together, it is found to be a powerful general depressant, oxygenation being impaired, nervo-muscular activity reduced, the heart weakened, and the waste of the body increased through all the channels of excretion, and by loss of heat. When a full dose (1 to 3 gr.) is given, and vomiting induced, this general depression may threaten to pass into collapse, with pallor and coldness of the surface, and marked fall of the body temperature. On this account tartarated antimony may sometimes be employed with benefit as an **antipyretic** or febrifuge at the commencement of acute febrile attacks in sound robust subjects, more especially bronchitis, where the attendant increase of the bronchial secretion will be serviceable, and the possible emesis by no means contra-indicated. Great caution must, however, be exercised in prescribing this powerful depressant, and the best method of administering it is in very small doses in water every fifteen or thirty minutes, until the skin becomes moist and cool, when it may be stopped.

The unquestionable value of Plummer's Pill would appear to be partly referable in the same way to the action of antimony not only on nutrition, but on the various organs of elimination, including the skin and the kidneys.

BISMUTHUM.   Bismuth.   Bi. 210.

A crystalline metal; as met with in commerce it is generally impure.

From Bismuthum is made .

Bismuthum Purificatum,—Purified Bismuth.

Source.—Made by heating Bismuth with Nitrate of Potash.

Characters.—A crystalline metal of a greyish-white colour, with a roseate tinge.

Impurity.—Copper; giving coloured reactions.

From Bismuthum Purificatum are made .

a. **Bismuthi Subnitras.**—Subnitrate of Bismuth.   White Bismuth. $BiONO_3H_2O$.

Source.—Made (1) by dissolving Purified Bismuth in Nitric Acid; and (2) pouring the product into Water.   (1) $Bi_2 + 8HNO_3 = 2(Bi3NO_3) + 2NO + 4H_2O$.   (2) $Bi3NO_3 + H_2O = BiONO_3 + 2HNO_3$.

Characters.—A heavy white powder, in minute crystalline scales; insoluble in water.

Impurities.—Carbonate of lead; giving precipitate with H_2SO_4 when dissolved in HNO_3; arsenic; and chlorides.

Dose.—5 to 20 gr.

Preparation.

i. Trochisci Bismuthi.—2 gr. Subnitrate of Bismuth in each, Carbonate of Lime, and the usual ingredients of a lozenge. *Dose,* 1 to 6.

From Bismuthi Subnitras is made :

ii. **Bismuthi Oxidum.**—Oxide of Bismuth. Bi_2O_3.

Source.—Made by boiling Subnitrate of Bismuth in Solution of Soda.

Characters.—A dull lemon-yellow powder; insoluble in water, soluble in nitric acid mixed with half its volume of water.

Impurities.—As of the subnitrate.
Dose.—5 to 15 gr.

β. Liquor Bismuthi et Ammoniæ Citratis.

Source.—Made by dissolving Purified Bismuth in Diluted Nitric Acid, adding Citric Acid, and re-dissolving the precipitate with Ammonia, as it forms.

Characters.—A colourless solution, with a saline and slightly metallic taste; neutral or slightly alkaline to test-paper; mixes with water without change. 1 fl.dr. contains 3 gr. of oxide of bismuth.

Dose.—$\frac{1}{2}$ to 1 fl.dr.

γ. Bismuthi Carbonas.—Carbonate of Bismuth.

$2(Bi_2CO_5)H_2O$; an oxycarbonate.

Source.—Made by (1) dissolving Purified Bismuth in Nitric Acid and Water; and (2) precipitating by a solution of Carbonate of Ammonia. (1) $Bi_2 +$ $8HNO_3 = 2(Bi3NO_3) + 2NO + 4H_2O.$ (2) $4(Bi3NO_3) + 3(N_4H_{16}C_3O_8) = 2Bi_2CO_5 + 7CO_2 + 12NH_4NO_3.$

Characters.—A white powder, insoluble in water; soluble with effervescence in nitric acid.

Impurities.—The subnitrate, and its impurities.

Dose.—5 to 20 gr.

ACTION AND USES.

1. IMMEDIATE LOCAL ACTION AND USES.

Externally applied in the form of powder or ointment, bismuth acts only physically on the unbroken skin, protecting it from the irritation of cold and dirt. If the surface be inflamed, as in chapped hands, chapped nipples, irritable ulcers, and eczema, it is a mild **sedative** and **astringent**, soothing and drying up the part. Accessible mucous membranes are similarly affected by bismuth, when in a condition of catarrh: thus it is used with success as a "snuff" for nasal catarrh; as an injection in gonorrhœa and leucorrhœa; and in irritability of the cervix uteri as a pessary. Bismuth is not known to be absorbed from the surface.

Internally, the local action and uses of the subnitrate of bismuth constitute all, or nearly all, that is definitely known respecting it as a remedy. In the stomach it is insoluble, and exerts the same **sedative** and **astringent** action as on the skin, whether by affecting the nerves and local circulation, or by its mechanical properties, that is, by coating and protecting the mucous surface. Little or no good is to be expected from less than 20 gr. doses of the subnitrate to an adult, and these may be trebled with perfect safety. Bismuth is extensively used in this country in the treatment of pain and vomiting due to catarrh or organic disease of the stomach, such as the gastric catarrh that follows a surfeit of food or alcoholic excess, recurrent gastric ulcer, and cancer; also in some cases of so-called nervous or reflex vomiting, as in pregnancy and hysteria, where a true catarrh is often present. Bismuth may be given alone in such conditions, but is better combined, on the one hand, with alkalies, such as bicarbonate of soda, if there be much actual catarrh; or, on the other hand, with opium, if pain be the chief symptom. A combination of the subnitrate of bismuth and a variable number of grains of Pulvis Ipecacuanhæ Compositus is almost a specific for the pain and vomiting of ulcer and malignant disease.

The astringent and sedative influence of bismuth on the intestines constitutes it a valuable remedy for diarrhœa in delicate persons, such as children, phthisical subjects, and those who have been exhausted by other causes. In lienteric diarrhœa, probably referable to duodenal catarrh, it is sometimes invaluable. But in the intestines, as in the stomach, the addition of opium, in however small quantity almost, greatly assists its action, and in persistent cases of diarrhœa is absolutely necessary. The same combination with Dover's powder gives excellent results. Bismuth subnitrate is partly converted into the sulphide in the bowel, which imparts a characteristic leaden-grey colour to the fæces.

2. ACTION IN THE BLOOD.

Neither the insoluble nor the soluble (but weak) preparations of bismuth enter the blood in any quantity. Still, the metal has been detected, both here and in the tissues.

3. SPECIFIC ACTION.

Bismuth finds its way, but very slowly, through all the organs and tissues; but no specific effect can be traced to its presence, even when it is given in doses of several drachms. The so-called effects of bismuth, of the older authorities, were certainly caused by arsenic combined with it as an impurity.

4. REMOTE LOCAL ACTION.

Bismuth has been found in the urine, and it is said, in the milk. No use is made of its remote influence, if any such exist.

CHLORUM. Chlorine. Cl. 35·5.

Although not contained in the Pharmacopœia as the pure gas under its own name, chlorine is furnished by several important preparations, as follows :

a. **Liquor Chlori.**—Solution of Chlorine. Chlorine gas dissolved in Water.

Source.—Made by heating Hydrochloric Acid in water with Black Oxide of Manganese, passing the gas into Water, and shaking till it is absorbed.

Characters.—A yellowish - green liquid smelling strongly of chlorine.

Impurities.—Salts, not volatile; deficient Cl, detected volumetrically by hyposulphite of soda.

Incompatibles.—Salts of lead and silver.

Dose.—10 to 20 min. in water.

b. **Calx Chlorata.**—*See* page 50.

a. Liquor Calcis Chloratæ.—*See* page 50.

β. Vapor Chlori.—Chlorinated Lime and Water. —*See* page 50.

c. **Liquor Sodæ Chloratæ.**—*See* page 38.

a. Cataplasma Sodæ Chloratæ.—*See* page 38.

ACTION AND USES.

1. IMMEDIATE LOCAL ACTION AND USES.

Externally, the action and uses of chlorine depend upon the great affinity which it possesses for hydrogen, and its consequent power to decompose compounds in which hydrogen forms part of the molecule, such as ammonia, sulphuretted hydrogen, sulphide of ammonium, and water. The properties of the body on which it acts (chemical, vital, or both) are completely altered; whilst nascent oxygen is set free, and chlorine further combines with the remaining elements of the broken-down molecule. Thus it is a powerful **irritant** to the skin, causing redness, vesication, even sloughing, and coagulating the albuminates of the part. For the same reason chlorine is the most powerful of all disinfectants, deodorisers, and decolorisers, its activity as a disinfectant greatly exceeding that of carbolic acid, and even corrosive sublimate. As a stimulant and disinfectant, chlorine water, or the solutions of chlorinated lime or of chlorinated soda, may be applied to foul ulcers, dissection and poisoned wounds, diphtheritic surfaces; or used in contagious ophthalmia, ozœna, and other foul discharges from surfaces or cavities. Of much more extensive application is the disinfectant action of chlorinated lime and its preparations, apart from the body: to purify rooms, wash infected clothes, flush drains, and throw upon the stools of typhoid fever and cholera before they are disposed of.

Internally, chlorine exerts the same local action upon the parts with which it comes in contact; and is employed as a wash or gargle, to disinfect and stimulate foul ulcers of the mouth, tongue, and throat, especially in diphtheria.

In the stomach chlorine in dilute solutions becomes converted into hydrochloric acid and chlorides, and loses all further effect upon the body as the uncombined element.

Inhaled as the vapour, chlorine causes local irritation of the respiratory passages, with distressing pain in the throat and chest, spasm, cough, lachrymation, sneezing, and headache. It cannot be recommended in this form or for this purpose.

2. ACTION IN THE BLOOD, SPECIFIC ACTION, AND REMOTE LOCAL ACTION.

It is doubtful whether chlorine enters the circulation or reaches the tissues, uncombined; more probably it is entirely converted into chlorides. From the analogy of its powerfully disinfectant and bleaching properties apart from the body, it has been given, as an "alterative and stimulant," in typhus, typhoid fever, small-pox, and other "putrescent" diseases, as well as in chronic dysentery, and liver disease of a malarial origin. There is little evidence in favour of continuing its use in these cases.

IODUM. IODINE. I. 127.

Under this head will be discussed both Iodine and Iodide of Potassium, the form in which the element is generally administered internally. Reference will also be made to the other officinal iodides.

Iodum.—Iodine. I.

Source.—A non-metallic element, obtained principally from Kelp, the ashes of sea-weed.

Characters.—Laminar crystals of a dark colour and lustre, and peculiar odour. Solubility, 1 in 7,000 of water, 1 in 12 of rectified spirit, 1 in 4 of ether, sparingly in glycerine, freely in a solution of iodide of potassium or chloride of sodium. Seldom given as pure iodine.

Impurities.—Iodide of cyanogen; subliming as colourless pungent prisms. Iron: not volatile. Water; as moisture. Deficient iodine; detected by hyposulphite of soda.

Incompatibles.—Ammonia, metallic salts, mineral acids, vegetable alkaloids.

Preparations.

a. **Linimentum Iodi.**—Iodine, 5; Iodide of Potassium, 2; Camphor, 1; Spirit, 40.

b. **Liquor Iodi.**—Iodine, 20 gr.; Iodide of Botassium, 30 gr.; Water, 1 oz.

c. **Tinctura Iodi.**—Iodine, $\frac{1}{2}$; Iodide of Potassium, $\frac{1}{4}$; Spirit, 20. *Dose*, 5 to 20 min.

From Tinctura Iodi is prepared

Vapor Iodi.—Tincture of Iodine, 1 fl.dr.; Water, 1 fl.oz.

d. **Unguentum Iodi.**—Iodine, 32 gr.; Iodide of Potassium, 32 gr.; Spirit, 1 fl.dr.; Lard, 2 oz.

From Iodum is made .

Potassii Iodidum. — See *Potassium*, for *source* and *characters.* *Dose*, 2 to 10 gr. or more.

Preparations.

α. **Linimentum Potassii Iodidi cum Sapone.**— Iodide of Potassium, 1$\frac{1}{2}$; Hard Soap, 1$\frac{1}{2}$; Glycerine, 1; Oil of Lemon, $\frac{1}{8}$; Water, 10.

β. **Unguentum Potassii Iodidi.** — Iodide of Potassium, 64 gr.; Carbonate of Potash, 4 gr.; Water, 1 fl.dr.; Lard, 1 oz.

γ. Also all preparations of Iodum.

Iodine is also used in the production of the Iodides of Cadmium, Ferrum, Hydrargyrum (2), Plumbum, and Sulphur.

Solutions of Iodine may be decolorised by Hyposulphite of Soda.

ACTION AND USES.

1. IMMEDIATE LOCAL ACTION AND USES.

Externally applied, iodine is a powerful **irritant** and **vesicant,** decomposing organic molecules, and entering into loose chemical combination with the albuminous constituents of the parts. At the same time it stains the epidermis of a deep brown, causes considerable pain; and is absorbed into the blood, partly by the skin and partly by the air of respiration in the form of vapour. It is also a powerful **antiseptic and disinfectant.**

The tincture, liniment, and ointment of iodine are extensively used as stimulants and disinfectants to foul, callous ulcers, much like nitrate of silver; as vegetable parasiticides in ringworm; and as counter-irritants in subacute or chronic inflammation of joints, periosteum, lymphatic glands, the pleura, and the lungs. In these instances the chief effect is doubtless stimulation, but a certain amount of the iodine is absorbed, and acts specifically, as will be presently described. Iodine in solution is injected into cysts, goîtres, hydrocele, etc., with much success.

Iodide of potassium applied to the unbroken skin is neither irritant nor capable of being absorbed, unless decomposed by the sweat. It is readily taken up from the exposed mucous membranes. How much specific value can be attached to the iodide liniment is doubtful.

Internally, the local action of free iodine is also irritant, and it is successfully applied to the gums in periosteal toothache. Inhaled into the respiratory passages, it gives rise to cough, sneezing, severe pain over the frontal sinuses, distressing pains in the chest, and dyspnoea. Combinations of iodine with creasote and various soothing volatile substances, such as chloroform and ether, have lately come into repute as continuous inhalations in the so-called " antiseptic " treatment of phthisis, bronchitis, and other forms of chronic lung disease.

In the stomach and bowels, although it is gradually converted into the iodide or iodate of sodium, the irritant effects of free iodine are continued, with abdominal pain, sickness, and diarrhoea as the result. The iodides of potassium and sodium have rarely this effect, and it is only in the form of a salt that iodine is now administered internally. Iodide of potassium is also decomposed in the stomach, the sodium salt and albuminate being formed from it.

2. ACTION ON THE BLOOD.

Iodine is freely absorbed into the blood from mucous surfaces, and the sodium iodide quickly enters from the alimentary canal. In the blood the element is at first combined with sodium; but this salt appears to be decomposed, the iodine for a time set free, some of the red corpuscles broken down (if the amount of iodine be large), and bloody effusions and bloody urine make their appearance. Such results are to be carefully avoided in practice; and, as far as we know, less degrees of the same cannot be usefully applied to therapeutical purposes, unless the tendency to coagulation of the blood be somewhat increased by it.

3. SPECIFIC ACTION AND USES.

The iodide of sodium and albuminous compounds pass from the blood into the tissues with remarkable rapidity, and may be found in all of them, especially the excreting organs and lymphatic glands, whilst they appear very scantily in the nervous centres. Almost as quickly the iodine leaves the tissues; and in thus passing rapidly through the protoplasm of the body, and sharing in its metabolism by combining (probably very loosely) with the albuminous molecules, it no doubt **accelerates tissue change.** As no increase of urea accompanies this effect, nor bodily wasting, the iodine must either spare the liver (which is the chief source of urea), or accelerate the metabolism of the plasma, rather than of the tissue elements themselves. (See *Metabolism*, Part III.) However this may be, the following are the principal directions in which iodine affects nutrition, and their applications:

(1) The *lymphatic glands* are reduced in size by iodine, which is extensively used for **scrofulous** and other chronic enlargements of the glands, whether applied locally as iodine, or internally as the iodides.

(2) Certain *poisons*, which have intimately associated themselves with the albuminous structures, are disengaged from this combination by iodine. Lead and mercury may be swept out of the tissues by iodide of potassium administered for plumbism and hydrargyrism respectively. The principal application, however, of iodine is in the treatment of syphilis. Either the poison of this disease is thus eliminated from the system, or iodine hastens the life and disappearance of the small-celled growth by which syphilis is characterised. It is specially valuable in the tertiary forms of syphilis, when mercury cannot be longer given with advantage; and nodes and other superficial enlargements, gummata in the viscera, and certain forms of skin disease may be very successfully treated by the potassium salt. The same precautions must be observed with respect to the general health, and especially the preservation of digestion in a course of iodide, as were laid down under the head of mercury.

(3) In subacute and **chronic inflammations** of various kinds, such as exudations or effusions in connection with the joints and serous cavities, and some forms of pulmonary consolidation, iodide of potassium may promote absorption by stimulating the

local nutrition. The local application of iodine "paint" is combined in such cases.

(4) *Scrofula* is benefited by iodine, especially when it affects the lymphatic glands, enlargements of which are treated by the liniment, by the ointments of the iodides of lead or cadmium, or by interstitial injections (rarely); internally by iodide of iron, or iodine mineral waters, such as the water of Woodhall. On the contrary, phthisis is rarely benefited by iodides, unless there be a syphilitic taint present.

(5) In chronic *rheumatism*, when debility is not a prominent symptom, in gonorrhœal rheumatism, and in the arthritis of syphilis, the iodide may be beneficial. In chronic arthritic gout it is probably useless, or even prejudicial.

The nervous system, respiratory centre, heart and vessels, and the body temperature are all unaffected by iodine; and the depressing effect on these of large doses of iodide of potassium is believed to be caused by the potassium. The remarkably useful effect of potassium in relieving or curing **aneurism** is due to the reduction of the blood pressure by the alkali, the coagulating effect of iodine on the blood, and the specific effect of iodine on the chronic inflammatory changes (often syphilitic) in the wall of the artery which have led to the dilatation.

4. REMOTE LOCAL ACTION AND USES.

Iodine is rapidly excreted, appearing in the urine, the mucous secretions generally, and specially in those of the air-passages, the perspiration, saliva, bile, and milk. Part of the sodium salt which reaches the excreting organs is thrown out unchanged, part is decomposed, and iodine is again set free to exert its local action remotely.

The diuretic effect of iodide of potassium is not marked unless large doses be given, and probably depends upon the alkali, not on the iodine. The latter may, however, have an alterative action upon the kidney, and the iodide may therefore be used in some forms of chronic Bright's disease, combined with other remedies.

The excretion of iodine by the mucous membrane of the respiratory tract is of most interest to the therapeutist. In certain subjects, and probably when iodide of potassium contains free iodine as an impurity, its exhibition produces a series of distressing symptoms known as "iodism," consisting of coryza, the watery discharge from the nose being sometimes profuse; sneezing; intense pain of a bursting character over the frontal sinuses, commonly called "headache;" swelling and redness of the gums, hard and soft palate and fauces, foulness of the tongue, and increase of the mucus of the mouth ; cough and frothy expectoration, and a sense of heat and rawness in the trachea and chest. The phenomena of irritation of the respiratory mucosa by the out-going iodine are therefore identical with those produced by the immediate action of iodine by inhalation, but in a minor degree. When the secretion is deficient, the mucous membrane of the bronchi swollen and dry, and cough useless and painful, iodide of potassium is thus a valuable **expectorant**, quickly inducing a flow of thin mucus, by establishing secretion, or by liquefying tenacious mucus which may be plugging or irritating the bronchi. It is, further, an indirect **antispasmodic**, given with great benefit in asthma and emphysema. The iodide of ethyl (non-officinal) inhaled as vapour may rapidly relieve the spasm of asthma. Iodide of potassium is sometimes given in other respiratory diseases, *e.g.* in pneumonia, if the consolidation threaten to persist.

In escaping by the skin the liberated iodine produces in certain individuals peculiar **eruptions**, generally papular or slightly vesicular, rarely purpuric. The value of the drug in tertiary syphilitic diseases of the skin no doubt depends partly on this influence.

5. ACTION AND USES OF THE SEVERAL PREPARATIONS CONTAINING IODINE.

1. *Cadmii Iodidum :* Unguentum Cadmii Iodidi. — The ointment only is employed, and combines the stimulant effects of the two elements. It is rubbed into the skin over enlarged glands, stiff joints, etc. See *Cadmium.*

2. *Ferri Iodidum :* Pilula Ferri Iodidi and Syrupus Ferri Iodidi combine the action of the two important elements, and are especially indicated and extensively employed when iodine has to be administered for a length of time to anæmic subjects. This is the form in which iodine is usually given in scrofula, the syrup being a favourite remedy for strumous children.

3. *Hydrargyri Iodidum Rubrum* possesses chiefly the action of the per-salts of mercury, and is used accordingly. See *Hydrargyrum.*

4. *Hydrargyri Iodidum Viride* is also a mercurial rather than an iodide in its action, and is employed in syphilis much like calomel.

5. *Sulphuris Iodidum* is now used externally only, and is believed to produce the combined effects of the two alteratives.

BROMUM. Br. 80. BROMINE.

A liquid non-metallic element.

Source.—Obtained from Bitter, and from some Saline Springs.

Characters.— A dark brownish-red very volatile liquid, with a strong disagreeable odour; solubility, 1 in 30 of water.

Impurity.—Iodine; detected by starch test.

Not given internally.

From Bromum are made :

    1. Ammonii Bromidum.—See *Ammonium.* Dose, 5 to 30 gr.

    2. Potassii Bromidum.—See *Potassium.* Dose, 5 to 30 gr.

ACTION AND USES.

1. IMMEDIATE LOCAL ACTION.

Externally bromine is a powerful **irritant and escharotic.** Its local use is confined to the treatment of cancer of the cervix uteri (1 in 5 parts of rectified spirit). The bromides have no such irritant action unless in highly concentrated solution; nor are they absorbed from the unbroken skin.

Internally, the local action of bromine resembles that of chlorine, the vapour being intensely irritant, and, indeed, irrespirable. It is never used in this way.

The bromides taken continuously for a time in full doses, or applied in strong solution to the throat, are said to reduce the sensibility of the fauces, so that the reflex movements of the parts, such as swallowing, vomiting, cough, etc., are not easily excited; and they may therefore be employed previous to important examinations or operations in connection with the larynx, or in excessive irritability of the parts. The bromides have but little effect of an irritant kind on the stomach or bowels, so that large doses (20 grains thrice a-day for years) may be readily borne. The greatest care must always be taken, however, to preserve the digestion and regularity of the bowels, in cases where bromides are continuously taken.

2. ACTION IN THE BLOOD.

Bromide of potassium enters the blood unchanged, where it is probably converted into the sodium salt by double decomposition with the chloride of sodium. For a moment it may be free in the blood, but no special action or therapeutic application can be referred to this circumstance.

3. SPÉCIFIC ACTION AND USES.

The bromides pass through the organs as such or as bromide of sodium, and have a very definite specific action upon them, which, speaking generally, is one of depression.

The *nervous system* is specially affected. **Loss of reflex excitability** in connection with all the sentient surfaces of the body follows the administration of full medicinal doses. This result is due partly to depression of the peripheral (sensory) nervous filaments, but chiefly to reduced activity of the nervous centres in the brain and cord. At the same time the motor nerves are also soothed, and the *muscular* power (which we may conveniently consider along with the nervous), is much weakened. The phenomena of this general nervo-muscular depression are as follows, beginning with the highest centres: (1) The bromides lessen mental activity, readiness to react to emotional stimuli, and sensibility and irritability of mind generally, thus inducing a condition of brain favourable to the advent of sleep. They are thus indirect hypnotics, not acting like opium and chloral, but so reducing the patient's sensibility of his surroundings, bodily condition, or circumstances, as to prevent distraction, and allow natural sleep to intervene. It is uncertain whether the bromides act upon the nerve cells directly, or upon the cerebral blood-vessels. The soothing and hypnotic effects of the bromides are very extensively employed in restlessness and sleeplessness from mental strain, whether emotional or intellectual, in the acute specific fevers when similar symptoms are urgent, in acute alcoholism, and in mania. In the three last conditions a certain amount of chloral or opium may be advantageously combined with the bromides. Bromide of lithium, the most active hypnotic of the bromides, will sometimes remove the insomnia of gout. The most important application of the soothing action of the bromides is in epilepsy, which is now almost exclusively treated with these salts, unless they be contra-indicated. Hysteria, infantile convulsions, whooping-cough, general "nervousness," hypochondriasis, and the low despondent condition so common in women with uterine irregularities, are also relieved by bromides, although not with the success obtained in epilepsy.

The great vital centres of the *medulla* are depressed by bromides. Respiration becomes slower and is weakened, whence possibly part of the value of the drug in whooping-cough. The heart is also slower and weakened in its action ; chiefly, however, by depression of its nervo-muscular substance, not of the cardiac centre. Bromides are of much service, therefore, in nervous disorder of the heart, especially in hysterical, dyspeptic, and alcoholic subjects. The direct effect of these drugs on the vessels is unsettled; as a whole, the tension is reduced.

The **spinal centres, and spinal nerves and muscles, are all depressed** by the bromides, the former so much so that the convulsions of strychnia poisoning cannot be induced, and the two drugs are so far physiological antagonists. In such a case and in tetanus the bromides may be given, but are neither rapid nor powerful enough to be trusted to alone.

The temperature is lowered by bromides, but not to an extent of much practical value.

The ovarian and uterine functions are quieted, and menor-rhagia relieved, by the same drugs.

4. REMOTE LOCAL ACTION AND USES.

The bromides appear in the secretions within a few minutes after their administration, being eliminated by the kidneys chiefly, by the salivary glands, mammæ, skin, and all mucous surfaces. In passing through these excreting organs, the bromides break up and set free bromine, which exerts a remote stimulant effect on the parts. The composition of the urine is irregularly disordered; but not in a manner that can be turned to therapeutical account. The **skin** is markedly affected, a characteristic acne-like **eruption** appearing, or other forms of skin disease, which are familiar in epileptics consuming large quantities of the drug. Cough is occasionally set up, and conjunctivitis may also occur. The interest to the therapeutist of all these remote effects of the bromides lies in their pre-vention, if possible, in cases where the drugs have to be steadily taken for an indefinite time, an end which may sometimes be secured by combining them with arsenic.

Acidum Hydrobromicum Dilutum.—Diluted Hydrobromic Acid (U.S.P.). (*Not Officinal.*) A liquid composed of 10 per cent. of absolute Hydrobromic Acid, HBr, and 90 per cent. of Water.

Source.—By decomposing Bromide of Potassium by Sul-phuric Acid, and distilling. $2KBr + H_2SO_4 = 2HBr + K_2SO_4$.

Characters.—A clear colourless liquid, odourless, with a strong acid taste and acid reaction.

Dose.—20 min. to 2 fl.dr.

ACTION AND USES.

Hydrobromic acid possesses many of the properties of the bromides, but is less useful than bromide of potassium. It is said to prevent the cerebral symptoms produced by quinine, which it readily dissolves, and the after-effects of morphia, if given with these drugs.

SULPHUR. Sulphur. S. 32.

An elementary body found native as virgin sulphur, also as sulphides of metals.

Sulphur Sublimatum.—Sublimed Sulphur.

Source.—Prepared from crude or rough sulphur by sublimation.

Characters.—A gritty powder, of a fine greenish-yellow colour, without taste or odour until heated. Insoluble in water, soluble in oils and turpentine with heat.

Impurities.—Sulphurous and sulphuric acid; acid to test-paper. Arsenic, detected by agitating with ammonia, and evaporating. Earthy matters.

Dose.—20 to 60 gr.

Preparations.

a. Confectio Sulphuris.—Sublimed Sulphur, 4; Acid Tartrate of Potash, 1; Syrup of Orange Peel, 4.

Dose.—60 to 120 gr.

b. Unguentum Sulphuris.—Sublimed Sulphur, 1; Benzoated Lard, 4.

From Sulphur Sublimatum are made :

c. Sulphur Præcipitatum.—Precipitated Sulphur, "Milk of Sulphur."

Source.—Made by (1) boiling Sublimed Sulphur with Slaked Lime in water; (2) precipitating the filtrate with Diluted Hydrochloric Acid, washing and drying.

(1) $6S_2 + 3CaH_2O_2 = 2CaS_5 + CaS_2H_2O_4 + 2H_2O.$

(2) $2CaS_5 + CaS_2H_2O_4 + 6HCl = 6S_2 + 3CaCl_2 + 4H_2O.$

Characters.—A greyish-yellow soft powder, free from grittiness and smell of SH_2.

Impurities.—Sulphate of lime; detected microscopically as crystals. Sulphuretted hydrogen; detected by odour.

Dose.—20 to 60 gr.

d. Potassa Sulphurata.—Sulphurated Potash. See *Potassium.*

Preparation.

Unguentum Potassæ Sulphurata.—1 in 15½.

e. **Sulphuris Iodidum.**—Iodide of Sulphur, SI.

Source. — Made by fusing Sublimed Sulphur with Iodine.

Characters.—Greyish-black crystalline pieces. Solubility, 1 in 60 of glycerine; insoluble in water.

Preparation.

Unguentum Sulphuris Iodidi.—1 in $15\frac{1}{2}$.

Sublimed Sulphur is also contained in Emplastrum Hydrargyri, and Emplastrum Ammoniaci cum Hydrargyro.

ACTION AND USES.

1. IMMEDIATE LOCAL ACTION AND USES.

Externally applied, sulphur has probably no local action of itself, but is partially converted, by contact with the acid products of the skin, into sulphuretted hydrogen and sulphides, which are energetic substances. Whether, therefore, rubbed on as ointment, worn in flannel, distributed over the surface by fumigation, or given as a natural or artificial bath of " sulphur waters," it is not sulphur, but its hydrogen compound, which possesses local therapeutical properties.

Sulphuretted hydrogen, when brought in contact with the skin in any of the forms just mentioned, is a **vascular stimulant and nervous sedative.** It is probably on this account that sulphur has long been regarded as useful in relieving the pains of chronic rheumatism; and as an **alterative** in certain kinds of skin disease such as acne, in which the ointment of the Potassa Sulphurata is especially valuable. The solution of the gas is also absorbed by the skin, and is extolled (in the form of baths) in lead and mercury poisoning, syphilis, and chronic enlargements of joints. The *rationale* of these effects will be discussed under the head of its specific action.

Sulphur and sulphurated potash destroy the *Acarus scabïei*, and are used in the treatment of itch.

Internally, sulphur has been locally applied to the throat in diphtheria, but with disappointing results.

In the stomach it remains unaltered, and passes as such into the intestines, where it acts as a purgative, possibly by increasing peristalsis, more probably by stimulating the glandular structures. Medicinal doses of milk of sulphur, the Confectio, or the German Pulvis Glycyrrhizæ Compositus, are simple **laxatives,** producing an easy soft stool with little or no pain. Sulphur waters, drunk freely at Harrogate, Moffat, and Strathpeffer in this country, at Aix-la-Chapelle, Challes, Aix-les-Bains, and the Pyrenees, on the continent of Europe, and at the Blue Lick, Alpena, Sharon, and other springs in the United States, have a similar but more powerful effect, producing considerable disturbance of the bowels, and depressing the portal circulation. Sulphur and sulphur waters are extensively used as purgatives in congestion of the rectum and liver, hæmorrhoids, and other diseases of the great bowel; and the waters and baths combined are powerful evacuants and alteratives in plethora, hepatic engorgement, and gravel.

Sulphur escapes in a great measure unabsorbed in the fæces, partly unchanged, partly as sulphides of hydrogen and the alkalies which it has encountered in the bowel, the activity of purgation varying indirectly with the degree of absorption.

2. ACTION ON THE BLOOD.

The amount of sulphur which enters the blood in the form of sulphides of hydrogen and the alkalies, under the use of sulphur or sulphur waters, is usually insignificant. When inhaled into the circulation, sulphuretted hydrogen is a powerful blood-poison, acting both on the red corpuscles and the serum; it reduces the oxyhæmoglobin of the former, and converts the carbonates and phosphates of the latter into sulphides, sulphites, and sulphates; but this subject is not of therapeutical interest.

3. SPECIFIC ACTION AND USES.

The hydrogen and alkaline sulphides pass into the tissues from the blood, and act chiefly upon the central nervous system. When in large quantity, they induce rapid failure of the nerve centres, especially those of respiration and circulation, the subject dying rather of asphyxia than from the poisonous influence on the blood just described. It is possible that the headache and nervous depression which attend the use of sulphur waters in some persons are minor degrees of these effects. It is possible also that sulphur and its compounds, possessing these powerful influences on the blood and tissues (which appear to be of the nature of arrest of oxydation), may modify nutrition to some extent even in medicinal doses, and thus possess **alterative** properties. In chronic rheumatism, syphilis, gout, and skin disease they have been much prescribed from time immemorial, especially at watering places. Sulphide of calcium has lately been found useful in scrofulous disease of bones.

4. REMOTE LOCAL ACTION AND USES.

It is under this head that we find the principal suggestions for the therapeutical employment of sulphur. The sulphides which we have traced through the blood and tissues are variously excreted. By the kidneys they pass out as sulphates, and it is said that one half of a dose of Sulphur Præcipitatum can be thus recovered from the urine, but only one-fifth of Sulphur Sublimatum. If in excess, part is also excreted as sulphides. No special use is made of these facts. By the skin they escape as sulphides, giving the characteristic foul odour to the perspiration, and somewhat increasing· its amount. Sulphur is used as a mild **cutaneous stimulant and diaphoretic**, and has always been regarded as a valuable internal remedy for many skin diseases, such as acne, chronic eczema, psoriasis, and syphilitic eruptions. Drinking the waters and taking the baths at sulphur springs probably act in this remote local way. Sulphide of calcium is specially useful in boils. The sulphides are also excreted by the bronchi and lungs, giving their odour to the breath ; sulphur was once much used as an **expectorant,** especially in chronic bronchitis with abundant expectoration and gouty or rheumatic associations.

The valuable effect of sulphur waters, taken internally and used as baths, in cases of chronic rheumatism, gout, skin disease, plethora, etc., is principally, if not entirely, to be accounted for by the immediate and remote local action of the sulphides—on the bowels and portal system, and on the kidneys, skin, and bronchi respectively. It is an important fact that sulphur is a **purgative alterative.**

CARBO. CARBON. C. 12.

Two kinds of carbon are officinal, namely, animal charcoal and wood charcoal.

1. Carbo Animalis.— Animal Charcoal. Bone Black.

Source.—Made by exposing bones to a red heat without the access of air.

Characters.—A black powder ; contains only 10 per cent. of pure carbon, the rest consisting chiefly of phosphate and carbonate of lime.

Preparation.

Carbo Animalis Purificatus.—Purified Animal Charcoal. Animal Charcoal from which the salts have been almost wholly removed.

Source.—Made by digesting Animal Charcoal in

Diluted Hydrochloric Acid, washing the undissolved part, and heating to redness in a closed crucible.

Characters.—A black powder, inodorous, and nearly tasteless.

Dose.—20 to 60 gr.

2. Carbo Ligni.—Wood Charcoal.

Source.—Wood charred by exposure to a red heat without access of air.

Characters.—Black, brittle, porous masses, without taste or smell, and retaining the texture of wood; contains about 2 per cent. vegetable ash.

Dose.—20 to 60 gr.

Preparation.

Cataplasma Carbonis.—Wood Charcoal, Linseed Meal, Bread Crumb, and boiling Water.

Charcoal is also used pharmaceutically as a decolorising agent, in the preparation of such drugs as morphia and atropia.

ACTION AND USES.

Externally.—Charcoal absorbs and condenses many gaseous bodies and vapours, as oxygen, carbonic acid, etc., and attracts the colouring, odoriferous, and sapid principles of many liquid substances, for example, litmus, bitters, wines, and decomposing liquids in general. It is used as a **deodorant** and **disinfectant** to absorb the foul emanations from cancerous and other discharges, ulcers, and wounds, being either hung around the bed in bags, or directly applied in dust, or as the poultice (a bad form.)

Internally.—Charcoal is locally used as a dentifrice. When taken into the stomach in sufficient bulk, either pure, or in the form of biscuits, it absorbs any gas and acrid products of indigestion which may be distending and distressing the organ, and is useful as a **carminative** in some forms of flatulent dyspepsia. Animal charcoal has been recommended by Dr. Garrod as an **antidote** in poisoning by opium, nux-vomica, aconite, and other organic poisons, which it attracts from their solutions in the stomach, and renders inert. It is doubtful, however, whether the absorptive action of charcoal can be retained in the bowel, or even in the stomach, after it has been thoroughly brought in contact with water. In the intestines it may possibly reduce flatulence, deodorise the fæces, and thus reduce the reflex peristaltic movements, and relieve diarrhœa.

Charcoal is entirely evacuated by the bowel and is not absorbed, so that it exerts no specific action on the body.

GROUP IV.

THE ACIDS.

The officinal acids may be classified as follows ·

1. Inorganic Acids. — Acidum Sulphuricum, A. Nitricum, A. Hydrochloricum, A. Nitrohydrochloricum Dilutum, A. Phosphoricum Dilutum, A. Sulphurosum, A. Arseniosum. Of these, Acidum Arseniosum is described under *Arsenic*, and Acidum Sulphurosum under its own heading.

2. Organic Acids.—Acidum Aceticum, A. Citricum, A. Tartaricum, A. Hydrocyanicum Dilutum, A. Carbolicum, A. Benzoicum, A. Gallicum, and A. Tannicum. Of the organic acids, the first three only will be discussed here ; the action and uses of the other substances being but little connected with their properties as acids.

ACIDUM SULPHURICUM, NITRICUM, HYDROCHLORICUM, NITROHYDROCHLORICUM DILUTUM, PHOSPHORICUM DILUTUM, ACETICUM, CITRICUM, AND TARTARICUM.

These substances all possess distinctly acid properties, that is, they neutralise alkalies, and turn blue litmus red.

Acidum Sulphuricum.—Sulphuric Acid. H_2SO_4 98 per cent., $= 79$ per cent. SO_3, in water.

Source.—Obtained by the combustion of Sulphur, and oxydation by nitrous fumes.

Characters. —A colourless, oily-looking, intensely acid liquid.

Impurities.—Nitric acid, lead, and arsenic, organic matter ; detected by colour.

Preparations.

1. **Acidum Sulphuricum Dilutum.**—1 to about 11 of Distilled Water. *Dose,* 5 to 30 min.

From *Acidum Sulphuricum Dilutum is prepared :*
Infusum Rosæ Acidum. 1 of Diluted Acid in 80.

2. **Acidum Sulphuricum Aromaticum.**—1 to about 13 of Spirit, with Cinnamon and Nutmeg. *Dose,* 5 to 30 min.

3. Many Sulphates and other preparations.

Acidum Nitricum.—Nitric Acid. HNO_3, 70 per cent. by weight, in Water.

Source.—Prepared from Nitre by distillation with Sulphuric Acid and Water.

Characters.—A colourless, intensely acid fuming liquid.

Impurities.—Sulphuric and hydrochloric acids; mineral matter; excess of water; peroxide of nitrogen; known by yellow fumes.

Preparations.

1. **Acidum Nitricum Dilutum.**—1 to fully 4 of Distilled Water.   *Dose,* 10 to 30 min.

2. **Acidum Nitrohydrochloricum Dilutum.**—3 to 25 of Distilled Water, with 4 of Acidum Hydrochloricum.   *Dose,* 5 to 20 min.

3. Many Nitrates and other preparations.

Acidum Hydrochloricum.—Hydrochloric Acid. HCl, 31·8 per cent. by weight, dissolved in Water.

Source.—Obtained by the action of Sulphuric Acid upon Chloride of Sodium, and the solution of the resulting fumes.

Characters.—A nearly colourless, very acid liquid, with pungent odour.

Impurities.—Sulphuric and sulphurous acids, arsenic, and water; detected by ordinary tests.

Preparations.

1. **Acidum Hydrochloricum Dilutum.**—1 to 2¼ of Distilled Water.   *Dose,* 10 to 30 min.

2. **Acidum Nitrohydrochloricum Dilutum.**—4 to 25 of Distilled Water, with 3 of Acidum Nitricum. *Dose,* 10 to 30 min.

3. Many Chlorides, and other preparations.

Acidum Phosphoricum Dilutum. — Diluted Phosphoric Acid. H_3PO_4 dissolved in Water $=$ 10 per cent. P_2O_5.

Source.—Made by distilling Phosphorus with Nitric Acid and Water, heating, and diluting.

Characters.—A colourless sour liquid, with a strongly acid reaction.

Impurities.—Arsenic and lead; detected by H_2S. Sulphuric, nitric, hydrochloric, and pyro- and meta-phosphoric acids; detected by usual tests.

Incompatibles.—Lime-water, calcareous salts, and carbonate of soda.

Dose.—20 to 30 min.

Diluted Phosphoric Acid is used in preparing Syrupus Ferri Phosphatis, and Ammoniæ Phosphas.

Acidum Aceticum.—Acetic acid.   Anhydrous Acetic Acid, $C_4H_6O_3$, 28 parts in Water.

Source.—Prepared from Wood by destructive distillation and purification.

Characters.—A colourless liquid, with a pungent odour and strong acid reaction.

Impurities.—Lead, copper; sulphuric, hydrochloric, and sulphurous acids.

Preparations.

1. **Acidum Aceticum Dilutum.**—1 volume to 7 of Water.   *Dose*, 1 to 2 fl.dr.

2. **Oxymel.**—Acetic Acid, 1; Water, 1; Honey, 8.

Dose.—3 to 6 dr.

Acetic Acid is also used in preparing Acetum Cantharidis, Acetum Scillæ, Extractum Colchici Aceticum, Linimentum Terebinthinæ Aceticum, Liquor Epispasticus, and many Acetates.

Acidum Aceticum Glaciale.—Glacial Acetic Acid. $C_4H_6O_3$, 84 per cent. in Water.

Source.—Made by distillation from Acetate of Soda and Sulphuric Acid.

Characters.—A colourless acid liquid, with a powerful acetic odour.

Impurities.—Sulphurous acid; and water.

Glacial Acetic Acid is used in preparing Acetum Cantharidis, and Mistura Creasoti.

Acetum.—Vinegar.

Source.—Prepared from Malt and Unmalted Grain by acetous fermentation.

Characters.—A brown-coloured acid liquid, with a characteristic odour.

Impurity.—Excess of sulphuric acid; detected volumetrically.

Dose.—1 fl.dr. and upwards.

Vinegar is used in preparing Emplastrum Cerati Saponis.

Acidum Citricum.—Citric Acid. $H_3(C_6H_5O_7)H_2O$.

Source.—Obtained from the juice of the Lemon (*Citrus Limonum*), or of the Lime (*Citrus Limetta*), by neutralising it with Chalk, decomposing the Citrate of Lime thus formed by Diluted Sulphuric Acid, purifying, and crystallising.

Characters.—Colourless crystals, in the form of right rhombic prisms. Very soluble in water. 17 gr. in $\frac{1}{2}$ fl.oz. of water make a solution resembling lemon jnice in strength and acidity, and exactly neutralise 25 gr. of Potassæ Bicarbonas, 20 gr. of Sodæ Bicarbonas, or 15 gr. of Ammoniæ Carbonas.

Impurities. — Copper, sulphuric acid, mineral matters. Tartaric acid, detected by precipitate with acetate of potash.

Dose.—10 to 30 gr.

Citric Acid is used in preparing Ferri et Ammoniæ Citras, Ferri et Quiniæ Citras, and Vinum Quiniæ.

Acidum Tartaricum.—Tartaric Acid. $H_2(C_4H_4O_6)$.

Source.—Obtained from Acid Tartrate of Potash by neutralising its solution with (1) Chalk and (2) Chloride of Calcium, (3) decomposing the Tartrate of Lime thus formed by Sulphuric Acid, and purifying. (1) $2KHC_4H_4O_6 + CaCO_3 = CaC_4H_4O_6 + K_2C_4H_4O_6 + H_2O + CO_2$. (2) $K_2C_4H_4O_6 + CaCl_2 = CaC_4H_4O_6 + 2KCl$. (3) $2CaC_4H_4O_6 + 2H_2SO_4 = 2H_2C_4H_4O_6 + 2CaSO_4$.

Characters.—Colourless oblique rhombic prisms, with a strongly acid taste, readily soluble in water. 20 gr. exactly neutralise 27 gr. of Potassæ Bicarbonas, 22 gr. of Sodæ Bicarbonas, or $15\frac{1}{2}$ gr. of Ammoniæ Carbonas.

Impurities.—Lead, oxalic acid, lime, mineral matter, acid tartrate of potash.

Dose.—10 to 30 gr.

Tartaric Acid is used in preparing Sodæ Citrotartras Effervescens.

Carbonic Acid.—Although not officinal as such,

carbonic acid gas is extensively used in medicine, being obtained from Bicarbonates and Carbonates, commonly of Soda, Potash, or Ammonia, by decomposition with Citric or Tartaric Acid. The process is known as *effervescence.* The reaction may be thus represented :

$$3(NaHCO_3) + (C_6H_5O_7)H_3 = Na_3(C_6H_5O_7) + 3CO_2 + 3(H_2O).$$

| Sodæ Bicarbonas | + | Acidum Citricum | = | Sodæ Citras | + | Carbonic Acid | + | Water. |

ACTIONS AND USES.

1. IMMEDIATE LOCAL ACTIONS AND USES.

Externally.—Acids are irritants, and some of them very powerful corrosives. The strong acids are used as **caustics :**

nitric acid to destroy chancres; acetic acid, warts; sulphuric acid, some forms of malignant growths. Very dilute watery solutions, sponged on the skin in fever, cool the surface by evaporation, and thus act as refrigerants; whilst watery solutions of sulphuric acid used in this way appear to constringe the tissues, and diminish the sweating of phthisis.

Internally.—In the dilute form, acids act directly upon the contents of the alimentary canal, and are used in the treatment of poisoning by alkalies. In every instance the free acids quickly unite with bases in the digestive tract, and form neutral salts. In the mouth they are stimulants and sialagogues: they relieve thirst, rouse the appetite, and aid digestion by increasing the flow of saliva and gastric juice, the citrates, tartrates, and acetates being chiefly used for this purpose as acid drinks and fruits of great variety, *e.g.* in fever. In the stomach hydrochloric acid increases the acidity of the gastric juice, and is given for this purpose during or after meals, as a powerful stomachic. Carbonic acid, introduced in effervescing wines and waters, has a grateful sedative action upon the gastric nerves; and in the form of champagne and effervescing mixtures is a most valuable remedy in the treatment of sickness with exhaustion. The other acids assist gastric digestion but to a very small, possibly useless, extent. Reaching the duodenum, acids increase the acidity of the chyme and stimulate the liver, pancreas, and intestinal muscles and glands. Dilute nitric and nitrohydrochloric acids, given at the end of meals, are therefore used as cholagogues in intestinal dyspepsia with hepatic torpidity, especially in tropical cases.

2. ACTIONS ON THE BLOOD AND THEIR USES.

Acids render the blood less alkaline (but never acid, even in poisonous doses), by combining with part of the alkali of the liquor sanguinis. No special use is made of this property. Phosphoric acid increases the phosphates in the red corpuscles, and is thus hæmatinic. The vegetable acids, when given as salts of the alkalies, have an important deoxydising effect on the blood. For example, citrate of potash becomes converted in the blood into carbonate of potash, carbonic acid, and water, a portion, however, of the citric acid always remaining unoxydised (see *Potassium*), thus: $2(K_3C_6H_5O_7) + O_{18}$ (in blood) $= 3(K_2CO_3) + 5H_2O, + 9CO_2$. Citrates, tartrates, and acetates of potash, soda, ammonia, etc., in the effervescing form, may therefore be used to set free in the blood the carbonates of the alkalies, which cannot be so conveniently or safely given in large doses by the stomach. The vegetable acids have been used in the treatment of scurvy, apparently with doubtful success; and in rheumatism, with equally questionable results, beyond their action on the mouth, skin, and kidneys.

3. SPECIFIC ACTIONS AND USES.

In the tissues and organs each of the acids exhibits a specific action of its own. *Sulphuric Acid* is an **astringent** to the bowels, skin, and blood-vessels, and is a valuable remedy for diarrhœa, profuse sweating, and hæmorrhage. *Nitric* and *Nitrohydrochloric Acids* are cholagogue, specifically as well as locally; *e.g.* when administered by means of a foot-bath (8 fl.oz. to one gallon of water), or of a compress wrung out of the solution and worn over the hepatic region. Tropical enlargements of the liver may thus be reduced. The **tonic** influence of these acids is probably referable to their stimulating effect upon the gastric and biliary functions. *Hydrochloric Acid* enters the tissues as chlorides, and no specific action or use can therefore be credited to the small doses which can be given of it. *Phosphoric Acid* also possesses no further influence on the tissues than that of increasing *pro tanto* the amount of phosphates, and possibly the growth of bones; and its value in constitutional diseases is probably due to its action on the red corpuscles, and to the bases with which it is combined (iron, lime, etc.). As we have seen, *Acetic*, *Citric*, and *Tartaric Acids* never reach the tissues, being decomposed in the blood, unless given in large doses.

4. REMOTE LOCAL ACTIONS AND USES.

The acids, having chiefly entered into combination as neutral salts, or having been decomposed in the blood, produce remarkably little local action when they are escaping from the body in the secretions. *Sulphuric* acid is excreted chiefly by the kidneys, increasing very slightly the normal amount of sulphates; part escapes by the bowels as sulphates; part by the skin, this acid being **anhidrotic**. *Phosphoric* and *Hydrochloric* acids behave similarly. *Nitric* acid is believed to be partly decomposed into ammonia, and thus actually to diminish, to a slight degree, the acidity of the urine. *Acetic*, *Tartaric*, and *Citric* acids pass out of the body as carbonates, unless in excess, when they escape unchanged by the kidneys. The important point to be noted about all these acids, therefore, is, that they **do not, to any** considerable **or useful extent, increase the acidity of** the **urine.** It must be observed, however, that all the acids probably stimulate the kidneys and skin indirectly, by increasing the total amount of salts excreted by them.

Acidum Boricum.—Boric Acid. Boracic Acid. BO_3H_3. (*Not Official, except as a test.*)

Source.—Sodæ Biboras, by the action of sulphuric acid.

Characters.—In white glittering plates, odourless, with a slightly bitter taste. A weak acid. Solubility, 1 in 26 of cold water, 1 in 3 of hot water; freely in glycerine; less freely in spirit.

Dose.—5 to 30 gr.

Preparations.

1. "Boracic Lint." Lint steeped in a boiling saturated solution, and dried. Contains nearly half its weight of the acid.

2. Various Ointments and Lotions.

ACTION AND USES.

Externally, boracic acid destroys low organisms, a solution of 1 in 133 arresting the activity of bacteria, and it is thus an **antiseptic, disinfectant, and deodorant**. On the tissues it produces little or no irritation, and is thus peculiarly adapted for use as a surgical dressing. The lint is now extensively employed in the antiseptic system; and lotions, warm fomentations made from a boiling saturated solution, and ointments with paraffin and vaseline, have almost replaced for the time other applications to wounds and ulcers. As its action does not extend beyond the surface to which it is applied, boric acid is never used for dressing cavities. It relieves itching, in the form of a powder, ointment, or glycerine. It is a test for Rheum (q.v.).

Internally, boric acid is a gastro-intestinal irritant in large doses.

Acidum Sulphurosum.—Sulphurous Acid. Sulphurous acid gas, SO_2, dissolved in water, and constituting 9·2 per cent. by weight of the solution (rarely contains more than 5 per cent.).

Source.—Made by heating sulphuric acid with charcoal, and dissolving the gas in water. $4H_2SO_4 + C_2 = 4SO_2 + 2CO_2 + 4H_2O$.

Characters.—A colourless liquid, with a pungent sulphurous odour.

Impurities.—Sulphuric acid; mineral matters; excess of water, detected by volumetric starch and iodine test.

Dose.—½ to 1 fl.dr.

Non-officinal Preparations containing Sulphurous Acid.

Sodæ Sulphis.—Sulphite of Soda. $Na_2SO_3,7H_2O$.

Source.—Made by saturating a solution of carbonate of soda with sulphurous acid gas.

Characters.—White efflorescent prisms, with a sulphurous taste; feebly alkaline. Soluble in 4 parts of water.

Dose.—5 to 20 gr.

Sodæ Hyposulphis.—Hyposulphite of Soda. $Na_2H_2S_2O_4.4H_2O$.

Source.—Made by passing Sulphurous Acid gas into a solution of Carbonate of Soda, with Sulphur. $2Na_2CO_3 + S_2 + 2SO_2 + 2H_2O = 2(Na_2H_2S_2O_4) + 2CO_2$.

Characters.—Large colourless transparent crystals, odourless, with a cool, bitter, sulphurous taste. Soluble in $1\frac{1}{2}$ parts of water. The solution is an officinal test for I and Cl.

Dose.—10 to 60 gr.

ACTION AND USES.

1. IMMEDIATE LOCAL ACTION AND USES.

Sulphurous acid is a powerful **deoxydising** agent. Seizing on oxygen and water, it decomposes organic bodies, and at the same time produces upon them the **irritant** local effects of sulphuric acid, into which it is converted. It thus destroys low forms of living matter, including the organisms associated with fermentation, decomposition, and certain diseases, 1 part in 666 of water being sufficient for this purpose. Sulphurous acid is therefore applied to ringworms and foul wounds; and some kinds of sore throat are relieved by a spray of the officinal acid. Morbid fermentation in the stomach attended by the growth of organisms, such as *penicillium* and *sarcina*, may be quickly arrested by doses of min. 5 to min. 60 of the officinal acid; but the non-officinal salts are more convenient forms for internal use, being decomposed by the acids of the stomach. Sulphites given in full doses become converted into sulphates, and act as purgatives.

Sulphurous anhydride, although not officinal, is very extensively used for fumigating infected rooms and clothing, being probably the most powerful, certain, and convenient of all **disinfectants.** Sulphur is burned on a shovel or plate, the outlets from the room having been carefully closed, excepting the door through which retreat is made.

2. ACTION ON THE BLOOD, AND SPECIFIC ACTION AND USES.

Sulphites were once supposed to enter the blood and tissues, and to arrest morbid fermentation or fever processes within them. The evidence, however, is to the effect that sulphites are not absorbed as such, but as sulphates, and the benefit derived from them in fevers is probably due to the laxative and diuretic effects of the higher salts. The suggested decomposition of hyposulphites into sulphites and free sulphur, and their consequent alterative and disinfectant action in phthisis and other diseases, appear to be equally unreal.

3. REMOTE LOCAL ACTION.

Sulphites are excreted by the urine and bowels in the form of sulphates.

Acidum Nitrosum.—Nitrous Acid. HNO_2. (*Not Officinal.*)

This acid is not itself used in medicine, but the nitrites are active and valuable drugs. Those in use are nitrite of sodium and nitrite of amyl, as well as sweet spirit of nitre and nitro-glycerine. The sodium salt will be noticed here; the others under their own heads.

Sodii Nitris.—Nitrite of Sodium. $NaNO_2$. (*Not Officinal.*)

Source.—Made by heating nitrate of soda. $2NaNO_3 = 2NaNO_2 + O_2$.

Characters.—A white granular powder, deliquescent, with a cool saline taste, very soluble in water.

Impurities.—The nitrate, and caustic soda.

Dose.—2 to 10 grains.

ACTION AND USES.

Nitrite of sodium acts upon the blood, the heart, and vessels **like nitrite of amyl,** only less suddenly and markedly, and for a much longer period of time. It has been used with success in heart disease, with recurrent attacks of painful angina; less successfully in epilepsy. Its depressant action on the central nervous system is more marked than that of the amyl compound; and it also paralyses the peripheral nerves and the muscles not only in this way, but through the blood.

GROUP V.
WATER.

AQUA. WATER.

Natural water, the purest that can be obtained, cleared, if necessary, by filtration, and free from odour, taste, and visible impurity.

From Aqua is made :

Aqua Destillata.—H$_2$O. Pure water, obtained by distillation.

Preparations.

Very numerous throughout the whole Pharmacopœia.

ACTION AND USES.

1. IMMEDIATE LOCAL ACTION AND USES.

Externally, water acts chiefly as a means of applying heat or cold to the surface of the body, being readily obtained at any temperature that may be desired. To produce this effect, water may be applied in the form of baths of all kinds : Cold, cool, temperate, tepid, warm, hot, vapour, or variously medicated; also by sponging, douching, fomenting, etc. These subjects will be noticed in the third part of the work. Possessing these properties, water is used externally for purposes of cleanliness; for either raising or lowering the temperature of the body; relieving pain, insomnia, and delirium; removing spasms or convulsions; diminishing the circulation in deep parts by superficial " derivation," as in congestion of the brain, etc. Water is also used, in a purely local way, as a wash or dressing to wounds, and as the basis of warm fomentations in inflammations.

Internally, water is constantly being taken in the form of food and drink. It relieves thirst, improves digestion and intestinal action when drunk in moderation and at proper times, and in a purely physical way may reduce the local or general temperature of the body, for instance, as ice slowly sucked in sore throat and febrile conditions.

2. ACTION ON THE BLOOD.

Water is quickly incorporated with the circulating plasma. Great excess has been known to dissolve part of the red corpuscles, but this is a purely pathological effect.

3. SPECIFIC ACTION AND USES.

Water plays an essential part in tissue life and in the activity of all the organs. A copious supply of water increases nutrition up to a certain point, especially the deposit of fat, and is therefore extensively employed in hydro-therapeutics.

4. REMOTE LOCAL ACTION AND USES.

Water is excreted by the kidneys, skin, lungs, bowels — indeed, necessarily in every secretion. Increase of water in the urine is most readily induced when the skin is kept cool, and carries with it an excess of urea, phosphoric acid, and chloride of sodium. Water is thus a diuretic, and in one sense the most natural measure of the kind, being indicated when we desire simply to irrigate or flush the uriniferous tubules and urinary passages, and wash from them the products of disease, such as blood, leucocytes, cellular *débris*, and sediments. Some kinds of calculi may be dissolved by the steady consumption of distilled water, which carries away minute traces of the stone, whilst it prevents fresh accretion on the surface.

As a **diaphoretic**, water acts best when warm and combined with external heat. It is the basis of most familiar domestic measures for relieving feverishness by inducing perspiration, such as warm drinks of all kinds, and spirituous compounds.

GROUP VI.

THE CARBOHYDRATES AND OTHER CARBON COMPOUNDS.

ALCOHOL. ALCOHOL. C_2H_6O.

Besides the following preparations, which are commonly associated with alcohol, all the Tincturæ, Spiritûs, and Essentiæ, several of the Liquores, Linimenta, and Misturæ, and a few other compounds contain it in various proportions.

1. Alcohol.—Absolute Alcohol. C_2H_6O. Used only in chemical testing.

Source.—Made by shaking rectified spirit with carbonate of potash; decanting, and distilling with slaked lime.

Characters.—Colourless, free from empyreumatic odour. Sp. gr., 0·795.

Impurities.—Resins, or oil; detected by turbidity on dilution. Water; giving blue colour with anhydrous sulphate of copper.

2. Spiritus Rectificatus.—Rectified Spirit. Alcohol, C_2H_6O, with 16 per cent. of water.

Source.—Obtained by the distillation of fermented saccharine fluids.

Characters.—Colourless, transparent, with a pleasant odour, and strong spirituous burning taste. Specific gravity, 0·838.

Impurities.—Water; tested volumetrically. Amylic alcohol, beyond a trace; detected by excessive reduction of $AgNO_3$. Resin or oil; giving turbidity on dilution with water.

Preparation.

Spiritus Tenuior. — Proof Spirit. Alcohol with 51 per cent. of water. Made by mixing 5 parts of rectified spirit with 3 parts of water. Specific gravity, 0·920.

3. Spiritus Vini Gallici.—Brandy. Spirit distilled from French wine.

Characters.—A spirit of a light sherry colour, and peculiar flavour; contains about 53 per cent. of *alcohol*, with some *volatile oil*, and *œnanthic ether*.

Preparation.

Mistura Spiritus Vini Gallici.—"Brandy Mixture," "Egg Flip." Brandy and Cinnamon Water, of each 4 oz.; Yolks of two Eggs; Sugar, ½ oz. *Dose*, 1 to 2 fl.oz.

4. Vinum Xericum.—Sherry. A Spanish wine.

Characters.—Pale yellowish brown, containing 17 or 18 per cent. of *alcohol;* with *colouring matter, ethereal compounds, acid tartate of potash, malates, sugar, etc.*

Preparations.

The following Vina: Aloes, Antimoniale, Colchici, Ferri, Ipecacuanhæ, Opii, Rhei.

Vinum Aurantii is made by fermentation of a saccharine solution; Vinum Ferri Citratis and Vinum Quiniæ are made from Vinum Aurantii.

Amount of Alcohol (absolute, by weight) in the various substances containing it.

Absolute Alcohol.
Alcohol (U.S.P.), 91 per cent.
Spiritus Rectificatus, 84 per cent.
Alcohol Dilutum (U.S.P.), 45·5 per cent.
Spiritus Tenuior, 49 per cent.
Spiritus Vini Gallici (Brandy), about 39 to 47 per cent.
Spiritus Frumenti (Whisky), about 44 to 50 per cent.

Rum
Gin   } about 40 to 50 per cent.
Strong Liqueurs)

Port, Sherry, and Madeira, about 14 to 17 per cent.
Vinum Album Fortius (U.S.P.), about 11·5 to 14 per cent.
Vinum Album (U.S.P.), about 10 to 12 per cent.
Champagne, about 10 to 13 per cent.
Hock and Claret, about 8 to 11 per cent.
Beer and Cider, about 3, 5, or more per cent.
Koumiss (made from milk), about 1 to 3 per cent.

ACTION AND USES.

1. IMMEDIATE LOCAL ACTION AND USES.

Externally, alcohol is an **antiseptic** and **disinfectant**, employed as a constituent of lotions for ulcers and wounds. Applied to the unbroken skin, and the vapour allowed to escape, it is a powerful **refrigerant**, withdrawing heat from the body by its evaporation. In this form it is used to prevent or allay inflammations of superficial parts, such as the subcutaneous tissues, joints, and muscles; blanches the parts by vascular constriction; produces a sense of cold; and relieves pain, especially headache, due to vascular dilatation and throbbing. Spirituous lotions sponged on the skin also diminish the activity of the sweat-glands, and may be used in excessive perspiration as an anhidrotic. On the contrary, if the vapour be confined, and allowed to act upon the tissues underneath, or if the alcohol be rubbed into the part, it penetrates and hardens the epithelium, and irritates the nerves and vessels of the cutaneous structures, causing redness, heat, and pain, followed by local anæsthesia. In the form of brandy it is rubbed into the skin to prevent bed-sores, by hardening and disinfecting the epidermis. Spirituous liniments containing soaps, essential oils, and other **stimulants** (*e.g.* Linimentum Camphoræ and Linimentum Camphoræ Compositum), are applied with friction to increase the nutrition of parts which are the seat of chronic inflammation, induration, adhesions, stiffness, and pain, such as the joints and muscles in chronic rheumatism, periostitis, and paralysis, or to produce a rubefacient effect on a large area of skin, for instance, of the chest in bronchitis. Alcohol is absorbed by the unbroken skin.

Internally, the local action of alcohol begins in the mouth with its characteristic taste and a hot, painful, stimulating effect on the tongue and mucous membranes. If it be retained in contact with them, the epithelium becomes condensed and whitened, and the parts beneath anæsthetised. Some forms of toothache can thus be quickly and completely relieved, the spirit also acting as a disinfectant in the pulp cavity. Wines and other wholesome alcoholic liquids consumed during meals have an action of the first importance on the nerves of the tongue, palate, and nose. By virtue of their taste, flavour, and bouquet, they give a relish to food, increase the appetite, and stimulate the flow of saliva and the functions of the stomach.

In the stomach the action of alcohol is complex, and of great importance. (1) Alcohol mixes with the *contents* of the stomach; is partly decomposed into acetic acid; and precipitates some of the proteids of the gastric juice: so far it depresses digestion. (2) It stimulates the *mucous membrane*, dilating and filling the vessels with blood; excites and markedly increases the flow of gastric juice; sharpens the appetite; and renders the movements of the viscus more energetic: in these respects it greatly assists digestion. The total effect of a moderate dose of alcohol is decidedly to **favour** gastric digestion, especially in cases where the nerves, vessels, and glands lack vigour, as in old age and in the chronic dyspepsia of persons weakened by acute illness, town life, and anxious sedentary employments. Herein consists the value of a small amount of wine or wholesome ale taken with meat meals by such subjects. The danger lies in excess, which readily destroys the activity of the juice, and also sets up a secretion of alkaline mucus which greatly interferes with digestion—a common cause of acute dyspepsia.

(3) The action of alcohol on the gastric wall produces extensive effects of a *reflex* kind. The heart is stimulated by moderate doses, producing a pleasurable rise of pressure and sense of power. The vessels dilate universally, filling the active organs with blood, further increasing their activity, the brain being specially excited, and the skin flushed and warmed subjectively. If the quantity be large these salutary effects of alcohol as a **diffusible stimulant** may pass into depression; and the sudden ingestion of a large amount of spirit may prove rapidly fatal by shock. The reflex results of alcoholic stimulants, if properly applied, add to its value at meal times, by increasing the enjoyment of eating, and thus the digestive power. Certain forms of pain in the stomach and bowels are rapidly relieved by the local action of brandy, which also helps to expel flatus; and pain, spasm, irregular or feeble action of the heart, cold feelings of the surface, and low conditions of the brain are all quickly removed by the same reflex means, before the alcohol could be absorbed in quantity into the blood.

2. ACTION ON THE BLOOD.

Alcohol enters the blood unchanged, and is distributed by it to the tissues and organs, a small part only becoming lost in it as acetic and carbonic acid. The action of alcohol on the corpuscles is still obscure, but it probably binds the oxygen more firmly to the hæmoglobin, so that oxygenation of the tissues occurs less freely, and therefore less extensively. The effect of this upon metabolism will now be described.

3. SPECIFIC ACTION.

Alcohol is rapidly taken up by the various organs, chiefly unchanged. If given in moderate quantity, it is (1) completely oxydised in its passage through the tissues into carbonic acid and water, like other carbohydrates, that is, it is a food, or source of heat and energy. At the same time it produces two other equally important effects; for (2) it reduces the activity of metabolism or the oxydation of the tissues; and (3) it first stimulates, and afterwards depresses, the circulatory and nervous systems, quite independently of its action on tissue-change. These three effects of alcohol must be discussed separately.

(1) *Alcohol as a food.*—It may now be accepted as proved that, when taken in sufficiently small quantities, alcohol is oxydised in the tissues; and that it only passes out of the body unchanged, through the lungs, kidneys, etc., when so freely given that excretion occurs before oxydation has had time to take place. This decomposition of alcohol must necessarily develop vital force and heat, like the oxydation of sugar, fat, and albumen. Alcohol belongs to the class of foods which do not become an integral part of the living cells, or "tissue proteids," as does much of the albumen, salts, etc., but remain in the plasma which bathes the cells, are oxydised there, and constitute their pabulum, the materials which supply the active elements with much of their energy, the "circulating proteids," carbohydrates, etc. Thus it happens that alcohol can for a time sustain life when no food (so-called) is taken, as in confirmed drunkards, and in some cases of severe illness. Professor Binz, of Bonn, who has studied this question with great industry and success, has calculated how much energy is contained in a gramme of alcohol, and finds that two ounces of absolute alcohol yield about the same amount of warmth to the body as is supplied by an ounce and a half of cod-liver oil. The *uses* of alcohol as a food will be presently described along with its other applications.

(2) *Alcohol as a nutritive depressant.*—Whilst it is itself thus oxydised in the tissues, alcohol unquestionably interferes with the metabolism or oxydation of other substances, especially (it would appear) saving or sparing the wear and tear of the "tissue-proteids," or formed protoplasm of the cells. This has been determined from three facts observed in animals

supplied with alcohol; first, that less oxygen is absorbed; secondly, that the temperature falls, and the albuminous tissues, whilst they do not waste, tend to degenerate into fat, so that the body as a whole grows fat and gross; thirdly, and chiefly, that the amount of urea, uric acid, carbonic acid, and salts excreted, is decidedly diminished. These are settled facts; the explanation of them is more difficult. The interference of alcohol with the oxygenating function of the red corpuscle is one obvious cause of impaired metabolism; another is the extreme readiness of the alcohol when it reaches the tissues to seize upon the oxygen which is there, thus robbing as it were the fixed elements of their necessary share, and arresting their decomposition af the middle stage of fat. This remarkable property of alcohol of **saving tissue waste** is one of the foundations of its employment in fever, to be presently discussed.

(3) *Alcohol as a stimulant and narcotic.*—The *circulation* in every part of the body is stimulated by a moderate dose of alcohol. The rise in the force and frequency of the heart, and the dilatation of the peripheral blood-vessels, which together constitute this **increased circulatory activity,** are both so far reflex effects from the mucous membrane of the stomach, as we have already seen; but they are also in part direct, the alcohol exciting the nervo-muscular structures of the heart, the cardiac centre, possibly the vaso-dilator centres in the medulla and cord, and certainly the nervo-muscular tissue of the middle coat of the vessels. To these causes of circulatory excitement must be added the voluntary muscular movements which are much exaggerated under the influence of alcohol. When alcohol is taken in large quantities, its stimulant effect passes into depression, both reflex and direct, and death may result, in part at least from cardiac failure.

Upon the *nervous system*, the first effect of alcohol in moderate quantity is one of stimulation. The **nervous centres** are increased in vigour from the highest to the lowest, and in the same order of sequence. The imagination becomes brilliant, the feelings are exalted, the intellect is cleared, the will is strengthened, the senses become more acute, the feeling of bodily strength and ability is raised, and some of the appetites are temporarily excited. The centres of speech, and of muscular movements generally, are specially exalted, giving rise to animated talk and lively gesticulations; and, therewith, a sense of *bien être*, referable to the combined nervous and circulatory excitement, spreads over the system.

If the dose of alcohol be larger, these phenomena of stimulation are more pronounced, but very soon give place to

depression, which spreads, like the excitement, from the highest
to the lowest centres of the brain and cord. The intellectual,
emotional, and voluntary faculties become first inco-ordinated,
then dull, and finally completely arrested; the muscles are
first **ataxic** and next paralysed, so that after an unsteady, stag-
gering gait, the erect posture is impossible; and the consequent
depression of the respiratory and circulatory centres leads to
stertorous breathing, circulatory failure, and even death.
The effects of alcohol upon the nervous centres are referable
partly to dilatation of the blood-vessels of the brain and cord,
but certainly also to a direct action of the drug upon the nerve
cells.

The action of alcohol on the other bodily functions is
chiefly, if not entirely, indirect. Thus, the *muscles* are affected
solely through the nervous centres and nerves. *Respiration* is
first increased, then slowed and weakened, partly through the
special centre, but manifestly also, to a great extent, through
the muscles and the circulation. Death occurs partly by
asphyxia. The **bodily temperature is, on** the whole, lowered
by alcohol: (1) by increased circulation through the dilated
peripheral vessels; (2) by increased perspiration; (3) by diminished metabolism; and (4) after large non-medicinal doses,
by general depression. The sense of warmth is, on the contrary, increased by the flushing of the skin with blood, a condition which promotes bodily heat and comfort in a warm or
moderately cool atmosphere, but causes rapid refrigeration,
general vital depression, and even death, in low states of the
external temperature.

4. SPECIFIC USES.

The uses to which the complex specific action of alcohol
may be turned are many, and of great importance:

Alcohol is employed in **fever,** and other acute wasting
diseases, such as delirium tremens, and acute mania. The in-
dications in these conditions are to prevent or to make good
the great waste of tissues associated with the disease; to sustain
the heart and nervous system, which threaten to fail, as the
frequent pulse and the delirium testify; and to promote the
loss of heat, which is formed in excess, as indicated by the
thermometer, the dry-brown tongue, the sleeplessness, and the
general restlessness of the patient. We have seen that these
ends are all fulfilled to a certain extent by alcohol. When the
symptoms just mentioned appear, brandy or other form of
spirit, and wines of the stronger varieties, are given in a definite
amount per diem, according to the height of the fever, the
state of the pulse and heart sounds, the general strength, the

ability to consume food, the previous habits, and the age of the patient. It must be distinctly understood, however, that alcohol is by no means essential in every case of fever; the very opposite being the case. In delirium tremens (acute alcoholism), where food, in the ordinary sense of the word, can often be given only with the greatest difficulty, the very substance which, as a stimulant, has caused the disease, must be judiciously continued as a form of nourishment for a time.

In chronic diseases attended by great debility, want of appetite, and.possibly sickness, such as pulmonary phthisis, alcohol will also find its place as a true food.

As a **stimulant** the principal use of alcohol is in connection with the heart. This, as we have just seen, is an important part of its action in fever. Of all remedies in threatening death by cardiac failure (syncope, fainting), spirits are the best, being at once available, convenient, rapid in their action, and almost invariably successful, if recovery be possible. For this purpose, brandy, whisky, etc., should be given either pure or only slightly diluted, by the stomach, bowel, or even under the skin. Hardly less valuable is alcohol, given continuously in small regular doses, in chronic disease of the heart, when natural hypertrophy fails and dilatation ensues. Wine, rectified spirit, or various tinctures, may be given in such cases.

In *nervous* depression alcohol must be ordered with the greatest hesitation. In melancholia, or in despondency begotten by grief, anxiety, suspense, over-work, excess, and especially by indulgence in alcohol itself, this drug affords only too ready relief, as also in neuralgia, hysteria, and allied disorders and sleeplessness; and the recommendation of it by the practitioner is frequently abused, being employed as a pretext for continued intemperance. In such cases the best rule is to order a definite amount of some weak alcoholic drink, such as ale or claret, at meal times only; but even this recommendation is by no means always safe. Severe pain, such as neuralgia, is often successfully relieved on the same principle. Some forms of sleeplessness are readily overcome by warm alcoholic draughts at bed-time, or malt liquors; but here again great discrimination is requisite in ordering the remedy.

5. REMOTE LOCAL ACTION AND USES.

Alcohol given in medicinal doses is, as we have seen, almost entirely oxydised in the system, only 16 per cent. passing out unchanged, chiefly by the lungs, less by the kidneys, and least by the skin. This amount, however, includes ethereal and other complex bodies associated with alcohol in wines and spirits; and by far the greater part of the alcohol proper is excreted as carbonic acid and water.

The diuretic effects of spirits, wines, and especially gin and beer, are well known, and may sometimes be employed in medicine. The diaphoretic effect of alcohol and its applications have been already sufficiently discussed under fever.

Circumstances modifying the action and employment of alcohol. —The different alcoholic fluids act very differently, according to their strength; their other constituents, already enumerated; the presence of carbonic acid in them (sparkling drinks), which increases the rapidity of their action on the stomach and possibly of their absorption; the degree to which they are diluted with water; and the condition of the stomach as regards the presence of food. The age of the patient, the soundness of his kidneys and other eliminating organs, his habits as regards alcohol, and the amount of exercise which he can take, must also be carefully estimated in ordering the remedy. In conditions of waste and exhaustion, especially febrile states and after operations, large quantities (even 1 pint of brandy per diem) may sometimes be tolerated.

ALCOHOL AMYLICUM. AMYLIC ALCOHOL.

"Fousel Oil," $C_5H_{12}O$, with a small proportion of other spirituous substances.

Source.—Contained in the crude spirit produced by the fermentation of saccharine solutions with yeast, and separated in the rectification of such spirit.

Characters.—A colourless oily liquid, with a penetrating and oppressive odour, and burning taste; specific gravity, ·818· Boils at 270°. Sparingly soluble in water; freely in spirit and ether.

Impurities.—Other æthereal substances; detected by specific gravity and boiling point.

Preparation.

Sodæ Valerianas.—See *Valerianæ Radix*, page 272.

USE.

Amylic alcohol is used only to prepare Valerianate of Soda.

CHLOROFORMUM. CHLOROFORM. $CHCl_3$.

Terchloride of Formyl.

Source.—Made by (1 and 2) distilling Rectified Spirit with Chlorinated Lime and Slaked Lime (oxydising and chlorinating the alcohol); washing with Sulphuric Acid; and redistilling with Slaked Lime and Calcium Chloride.   (1) $2C_2H_6O + O_2 + Cl_{12} = 2C_2HCl_3O$ (chloral) $+ 6HCl + 2H_2O$.   (2) $2C_2HCl_3O + Ca2HO = 2CHCl_3 + Ca2CHO_2$ (formate of lime).

Characters.—A limpid, colourless, heavy, volatile liquid, of an agreeable ethereal odour and sweet taste.   Solubility, 10 in 7 of spirit; 1 in 200 of water; freely in olive oil and turpentine.   Sp. gr., 1·49.

Impurities.—Hydrocarbons; detected by green colour with sulphuric acid.   Non-volatile compounds; detected by residue and unpleasant odour after evaporation.   Alcohol; detected by opalescence when dropped into water.

Dose.—3 to 10 min.

Preparations.

1. **Aqua Chloroformi.**—1 in 200 of Water.   *Dose,* ½ to 2 fl.oz.

2. **Linimentum Chloroformi.**—1, with 1 of Camphor Liniment.

3. **Spiritus Chloroformi.**—"Chloric Ether."   1 to 19 of spirit.   *Dose,* 10 to 60 min.

4. **Tinctura Chloroformi Composita.**—2; Spirit, 8; Compound Tincture of Cardamoms, 10.   *Dose,* 20 to 60 min.

ACTION AND USES.

1. IMMEDIATE LOCAL ACTION AND USES.

Externally applied, and allowed to evaporate, chloroform causes a sense of coldness, and depresses the terminations of the sensory nerves of the part, thus reducing sensibility or removing pain.   If, on the contrary, the vapour be confined, or the chloroform rubbed into the skin, it acts as an irritant, causing redness and even vesication, with a sense of heat and pain, followed by anæsthesia of the part.   A similar effect is produced on all exposed mucous membranes.   **As a local anæs-thetic,** chloroform may be applied on lint, covered closely with a wine-glass (*e.g.* in temporal headache); or in the form of liniment with various combinations of belladonna and other anodynes, which are used for the relief of lumbago, neuralgia, etc.   The student must understand, however, that the local anodyne effect of chloroform bears a very inferior relation to its rapid and powerful action as a general anæsthetic.

Internally.—When it is given by the mouth, chloroform produces an intensely hot, sweet taste, which renders it useful in pharmacy to cover the nauseous, bitter, and astringent characters of many drugs. It may also be used to relieve toothache. Like alcohol, it causes reflex salivation, and in this way, as well as by a **carminative** action on the stomach, the compound tincture, spirit, and aqua are useful adjuvants to stomachic and tonic mixtures, relieving pain, vomiting, and flatulency. In full doses it may give rise to vomiting, as is frequently seen after anæsthesia. A few drops of chloroform inhaled from a sponge or piece of lint (quite apart from its action and use as a general anæsthetic), rapidly soothe the respiratory nerves, and may be employed to arrest spasm of the glottis, asthma, and spasmodic or dry useless cough attending irritation of the air passages.

2. ACTION IN THE BLOOD.

Chloroform enters the circulation by the respiratory organs, stomach, and the unbroken skin, as well as subcutaneously. Chiefly as chloroform, partly as various products, it mixes with the blood; but its action on the living circulating blood is still obscure.

3. SPECIFIC ACTION AND USES.

Chloroform reaches the tissues very rapidly, especially if administered in the form of vapour freely mixed with air, as it always is when given as a general anæsthetic. Its most important action is exerted upon the central nervous system, and demands detailed description. The phenomena which it produces will first be noted; secondly, an analysis will be made of these; thirdly, the uses of chloroform will be enumerated; and fourthly, the method of administering the anæsthetic, and certain necessary precautions will be briefly indicated.

1. **Phenomena of chloroform anæsthesia.**—*a. First stage.* The first effect of the inhalation of chloroform on the nervous system is powerful **stimulation**, but almost from the commencement this is accompanied by a certain amount of disorder. The first few inspirations seem to rouse the cerebrum to increased activity, an effect due to the direct action of the anæsthetic on the nerve-cells of the convolutions and partly, perhaps, to vascular disturbance. The highest centres are first and chiefly excited so that the imagination and feelings immediately become exalted, always, however, with some confusion. For a moment the senses may be quickened, but they are quickly disordered and depressed: vision, hearing, and touch become dulled or blurred, and a strange feeling of lightness, freedom, tingling, or numbness pervades the surface and the extremities. All these sensations are strictly central, probably convolutional in origin.

At the same moment, or almost immediately after, the chloroform rouses the spinal or muscular centres, and various gesticulations and spasmodic or struggling movements may ensue.

The medulla oblongata is next affected, the centres of circulation and respiration being stimulated, so that the pulse and respiration become more frequent (although the latter is more shallow), the face flushed, the blood pressure raised. At this point the skin becomes moist, a red rash in irregular patches may appear on the neck and chest; and the pupils may dilate slightly.

These phenomena vary greatly in different instances, with the constitution and condition of the nerve-centres, the temperament and habits of the individual, laughing or crying or noisy struggling being the most prominent feature in many cases.

b. Second stage.—The second effect of chloroform on the nerve-centres is depression. The same parts continue to be affected by the drug, but their functions, instead of being increased or simply disordered, are first diminished, and at last completely arrested. Consciousness now ceases, with the appearance of heavy sleep. Perception and sensation are annulled: the patient sees nothing, hears nothing, feels no pain. For the same reason, reflex excitability is first diminished and then lost: irritation of any part by tickling or pinching no longer induces movements of the limbs; at last, even touching the cornea causes no reflex rolling of the eye-ball or winking of the lids.

As the anæsthesia deepens, the reflex and automatic excitability of the cord and medulla is also diminished, and the phenomena that ensue affect all the parts supplied by these centres. The muscular tone is lost, and the voluntary muscles become paralysed and relaxed. The pupil is semi-contracted, and may dilate on stimulation of afferent nerves. The heart and respiratory organs are no longer excited, but their centres in the medulla being now depressed, their action is laboured, the pulse falling in frequency (a striking change from the previous acceleration) and in strength, and the respiratory movements being slow, heavy, and attended by noise or stertor.

Now is the time for the surgeon to operate, anæsthesia being complete, whilst the depression of the vital functions is still within safe limits. The effects may be expected to pass off in a few minutes if the administration be stopped; and although the amount required to complete the second stage varies greatly with the subject and other circumstances, it may be said that from 1 to 4 fluid drachms will probably have been given up to this point.

K—8

c. Third stage.—Beyond the second stage or degree, chloroform anæsthesia is highly dangerous, the further action of the drug being attended by complete **loss of all reflex excitability** of the cord and medulla. The sphincters relax, the pupils are widely dilated and fixed, the globes prominent. The respiratory centre is no longer irritable, and the movements of the chest become weaker, irregular, sighing, and finally cease. The cardiac centre fails, the heart beating irregularly and feebly, and at last stopping in diastole, both from central and from direct nervo-muscular depression. The blood-vessels dilate, the pressure falls to zero, and the circulation has come to a standstill. It is obvious that the direct effects of chloroform on the respiratory centre are complicated towards the last by venosity of the blood. Death may occur through the heart, the respiration, or both together.

2. **Analysis of** the phenomena **of chloroform anæsthesia.** —Chloroform anæsthesia affords us an excellent opportunity of studying the action of a drug upon the various centres of the nervous system, from the highest downwards. The first parts to be stimulated are the cerebral centres with mental functions, the control of the special senses and consciousness; and these are the first to be depressed and finally annulled. The lower cerebral and spinal centres are affected less and somewhat later, so that a certain degree of excitement of these accompanies the first cerebral depression; and the spinal centres being no longer controlled by the cerebral, irregular excessive movements of the limbs ensue. As the depression deepens in the spinal centres, the muscles are paralysed. Lastly, the lowest centres of all, those of organic life, connected with the heart, vessels, respiratory organs, and sphincters, situated in the medulla and cord, yield to the action of chloroform. Although affected from the first, it is not until the higher parts have become completely overpowered that the functions of these vital centres are seriously impaired, and death threatens. It is on account of the safe order of invasion of the different centres by chloroform that it has been selected as the proper agent for temporarily arresting consciousness; we shall find that many other powerful drugs equally depress the nervous system, but in a direction exactly the reverse.

The peripheral nerves are affected last of all in general anæsthesia, and it must be repeated that the loss of sensibility to the knife is due to a *central*, not a peripheral effect.

The muscles are finally affected directly, as well as through the nervous system. The pupil is dilated in the first stage, probably by stimulation of the sympathetic; and contracted in the second, and dilated in the third stage, by stimulation and paralysis respectively of the third nerve or its cerebral centre. The other involuntary muscles are less obviously paralysed, and the parturient uterus contracts freely in complete anæsthesia, with some loss, however, of vigour and regularity.

3. **Specific uses of chloroform.**—The circumstances under which chloroform anæsthesia may be employed are the following: (1) In *operations attended by pain.* These need not be particularised. (2) In *operations where muscular action or spasm has to be overcome :* reduction of herniæ, dislocations and fractures ; catheterism. (3) In *diagnostic manipulations :* exploration of the abdomen externally and *per rectum.* (4) In *diseases attended by excessive pain*, especially biliary and renal calculus. (5) In *parturition*, in certain subjects and conditions, the degree of anæsthesia induced being generally slight until the moment of birth. (6) In *spasmodic diseases*, such as tetanus, hydrophobia, uræmia, puerperal convulsions, the *status epilepticus*, severe chorea, and hiccup.

4. **Method of administration, and principal precautions to be** observed **in chloroform anæsthesia.**—This is a purely practical subject, to be learned by experience and not in theory. The student has frequent opportunities of witnessing the administration of anæsthetics by skilled persons, and he must closely and carefully observe every effect of the chloroform upon the patient. He will do well to interpret every phenomenon as it arises, such as mental and muscular excitement, the character of the breathing, the colour of the countenance, and (if possible) the state of the pulse, into exact physiological terms, as explained above; as, for example, stimulation of the convolutions and cord, interference with the respiratory centre, etc. He will thus come to appreciate accurately the condition of the patient at any moment, and be prepared to assist in administering anæsthetics himself. Many purely practical points will then have to be learned : the selection of suitable cases for anæsthesia; the preparation of the patient; the choice of the anæsthetic and of an inhaler; the position of the patient; the method of watching the face, eyes, pulse and respiration ; the detection of unfavourable symptoms, and their immediate treatment ; and, finally, the after treatment of the case. All these and other matters connected with the administration of anæsthetics can be but briefly referred to in the following paragraphs :

a. Selection of cases.—Chloroform must be given with great caution to the aged and infirm, to persons who are subject to attacks of faintness, or known to suffer from fatty degeneration or dilatation of the heart, to very fat, and very anæmic persons, to epileptics, to chronic drunkards, to the subjects of extensive disease of the lungs or respiratory passages. Nitrous oxide gas or ether must be preferred in such subjects, according to the length of the operation. Valvular disease of the heart with compensation suggests special care, but is not a contra-indication. Operations on the mouth, nose, throat, attended by possible bleeding into the glottis, demand special precautions, whether by great expedition, special postures of the patient, or even previous tracheotomy. It must never be forgotten, however, that when an operation is absolutely necessary, it can always be more safely performed with anæsthetics than without their aid; and that before the days of ether and chloroform, many persons died under an operation, from fear, faintness, and shock, the danger from which is completely removed or greatly diminished by anæsthetics.

b. Preparation of the patient.—Insensibility is more rapid when the stomach is empty. No solid food should therefore be given for at least six hours before the operation, which should, if possible, be performed early in the morning when digestion has been completed and the anæsthetic is rapidly absorbed. If the patient feel faint under these circumstances, a small quantity of brandy and water may be given before operation. Artificial teeth must be removed. The respiration and pulse should be carefully noted before commencing inhalation.

c. Selection of the anæsthetic: purity of the same.—The anæsthetic agents in general use at the present time are chloroform, bichloride of methylene, ether, and nitrous oxide gas. Of these, ether and nitrous oxide are unquestionably to be preferred, unless there be some special reason to the contrary. The purity of the drug is best insured by purchasing it from well-established makers, and not by attempting to test it for oneself; and the same manufacture should always be used, if possible. It may be advisable to commence with one anæsthetic, and then, as circumstances alter during the operation, to change it for another.

d. Selection of the apparatus.—This will depend on circumstances and on the taste and experience of the administrator. Whilst elaborate inhalers are used in hospitals, it is satisfactory to know that the simplest apparatus is equally safe, such as a handkerchief or towel made into a cone, care being taken that chloroform vapour is mixed very freely with air, but that with ether, on the contrary, the atmosphere is excluded as completely as possible. A few capsules of nitrite of amyl and a straight polypus forceps should be ready at hand.

e. Position of the patient.—The administrator must accommodate himself to the convenience of the operator, whose eye and hand must never be interfered with. If possible, the patient's head should be placed in such a position on the edge of a pillow that the saliva may flow from the mouth instead of into the stomach, and that the tongue may not fall back and produce dyspnœa. It is essential that the patient's chest and abdomen should not be compressed in the slightest degree by clothes or by the arms of the assistants, or confined by bandages. The most comfortable position for the patient is on the side, with one hand and fore-arm beneath the pillow ; and as a rule it is better to induce insensibility in this position, and afterwards arrange the patient for the surgeon, than to ænasthetise him in the constrained attitude often required in operations.

f. Administration.—The confidence of the patient should first be gained by a few minutes' conversation, whilst he is reassured as to the result and instructed how to breathe. When inhalation has commenced, the administrator must not even for a single instant cease to watch the face, respiration, and pulse. The degree of insensibility necessary for different cases varies greatly, the least being required for uterine, the most for rectal operations. The loss of the corneal reflex, and stertorous breathing, are generally employed as tests of insensibility, but no single sign can be relied upon. The smallest possible quantity of the drug should always be given ; and patients once thoroughly anæsthetised by ether may be kept under its influence for many minutes by rebreathing the air of expiration loaded with its vapour mixed with some fresh air.

g. Complications and unfavourable symptoms.—*Vomiting* is generally preceded by pallor of the face or a few deep inspirations. When it occurs, care must be taken that nothing is drawn into the larynx ; the head should therefore be thrown forward, and the mouth opened by pressure on the symphysis of the jaw, or by inserting a pair of forceps between the teeth. Should vomited matter be inhaled into the respiratory passages and asphyxia threaten, laryngotomy must be immediately performed.

Lividity of the face and prolonged deep *stertor* should be checked by raising the shoulders so that the diaphragm may descend more easily, and by making the patient breathe fresh air. The position of the head is to be changed until respiration is more easy; the vessels of the head and neck must be allowed to empty themselves well and quickly ; and the mouth may have to be opened to its fullest extent, which induces a deep inspiration, the following expiratory effort often clearing the larynx and fauces of tenacious mucus which had been obstructing the entrance of air.

Pallor of the face is to be combated by lowering the head and shoulders ; if severe, by dropping the head over the end of the table. If this should not succeed, the vapour of nitrite of amyl should be given.

Shallow breathing, especially if intermittent, should be anxiously watched; and if it increase, artificial respiration should be at once resorted to, on no account waiting for the respiration to cease.

h. After-treatment.—Absolute quiet and keeping the eyes closed often prevent sickness after an operation. The whole surface of the body being carefully covered to prevent chill, the room should be cleared of ether vapour as quickly as possible. Cough induced by ether is often attended by blood-stained mucus, which, with these precautions, is of no consequence. Food should not be given within two hours after the operation, and for the first twelve hours should be entirely cold, and consist chiefly of soups and jellies, milk being avoided. A teaspoonful of burned brandy will often relieve the after-sickness, when all other measures have failed.

4. REMOTE LOCAL ACTION.

Chloroform is excreted in part, as such, by the kidneys, lungs, and skin; part is lost in the system. No use is made of its remote effects, although small doses given by the mouth are said to increase all the secretions.

ÆTHER. Ether. $C_4H_{10}O$.

A volatile liquid prepared from Alcohol, and containing at least 92 per cent. pure ether. $C_4H_{10}O$.

Source.—Made by (1 and 2) distilling Rectified Spirit with Sulphuric Acid, purifying with Slaked Lime and Chloride of Calcium, and redistilling. (1) $C_2H_6O + H_2SO_4 = C_2H_6SO_4$ (sulphovinic acid) $+ H_2O$. (2) $C_2H_6SO_4 + C_2H_6O = C_4H_{10}O + H_2SO_4$.

Characters.—A colourless, very volatile liquid, with peculiar strong odour and hot taste. It is entirely dissipated in vapour when exposed to the air, boils below 105° Fahr., and is very inflammable. It contains 8 per cent. of spirit. Specific gravity, 0·735.

Impurities.—Alcohol ; tested by specific gravity.

Dose.—20 to 60 min.

Preparations.

1. **Æther Purus.**—Pure Ether. Ether free from alcohol and water.

Source.—Made by shaking ether with water, separating, purifying by quicklime and chloride of calcium, and distilling.

Characters.—Specific gravity, 0·720. Boils at 96° Fahr. Given by inhalation.

Impurities.—Alcohol and water; detected by specific gravity.

2. **Spiritus Ætheris.**—Ether, 1; Rectified Spirit, 2. Specific gravity, 0·809. *Dose*, 30 to 60 min.

From Spiritus Ætheris is prepared:

Tinctura Lobeliæ Ætherea. See *Lobelia.*

Ether is also used in making Collodion and Liquor Epispasticus; and in many pharmaceutical processes.

ACTION AND USES.

1. IMMEDIATE LOCAL ACTION AND USES.

Externally.—When allowed to evaporate, ether is a powerful **refrigerant** and **anæsthetic**, abstracting heat and depressing the nerves of the part. It is used in the form of Dr. Richardson's spray to relieve the internal local pain of neuralgia, and more frequently to prevent pain in minor surgical operations, the parts being completely frozen in the course of a few minutes by the spray of pure ether from a proper apparatus. If the vapour be confined, or the ether rubbed into the skin, a **rubefacient** or even **vesicant** effect is produced, as with chloroform. The Ethereal Extract of Mezereon and Liquor Epispasticus are powerful vesicants and **counter-irritants**.

Internally.—Ether has a powerfully burning, disagreeable taste, and causes local irritation and reflex salivation in the mouth, like chloroform. Reaching the stomach, it acts as a local stimulant to the blood-vessels, nerves, and muscular coat, and is therefore used as a **carminative**, relieving pain and sickness, and expelling flatulence, especially in nervous subjects. At the same time, it acts reflexly from the gastric mucosa upon the heart and respiratory organs, as a powerful systemic stimulant. It is a very useful ingredient of powerful **anti-spasmodic** draughts, as will be presently described. Given with cod-liver oil, it renders it more palatable to some patients, and more digestible, possibly by stimulating the pancreas.

2. ACTION ON THE BLOOD.

Ether is absorbed into the blood with remarkable rapidity, and acts here like chloroform.

3. SPECIFIC ACTION AND USES.

The specific action of ether and its employment as an anæsthetic so closely agree with those of chloroform that the reader is referred to the description of the latter drug, and the differences between the two substances only require to be mentioned here.   These are :

1. Ether must be administered nearly pure, say 70 per cent. of the vapour with 30 per cent. of air; whilst but 3 to 4 per cent. of chloroform is given, with 97 or 96 per cent. of air.

2. With ether the stage of stimulation is more protracted; there is more struggling; and the stage of anæsthesia is shorter and the degree less profound.   Ether is therefore said to be safer than chloroform.

3. Ether depresses the heart and vessels less than chloroform, the heart continuing to beat after respiration has been arrested by an excessive dose.   The respiratory centre is also less depressed.   For these reasons, also, ether is called a safe anæsthetic.

4. Ether has a much less pleasant smell than chloroform.

In choosing between ether and chloroform, preference must be given to the safer **anæsthetic,** and the use of ether has accordingly been much revived during the last few years.   Under certain circumstances chloroform is preferable, as in operations about the mouth, ether causing a profuse secretion of ropy mucus; in operations where a light or the cautery might come into contact with the ether vapour and cause an explosion; in operations which must be hastily undertaken and completed; and in parturition, where profound anæsthesia is unnecessary. Infants bear chloroform well, and their delicate respiratory passages are less irritated by it than by the pungent vapour of ether.

Given by the stomach in small doses, ether increases the activity of the circulation and nervous system—partly, as we have seen, by reflex action from the gastric wall; and is used as a powerful and rapidly diffusible stimulant and antispasmodic.   It is given largely in cardiac failure, faints, angina pectoris, palpitation, and depression, being even more rapid in its effects than alcohol, but more evanescent, and of course less available in emergencies.   Its antispasmodic powers make it useful in hysterical and epileptic threatenings; and in spasmodic cough and asthma it is one of the most valuable remedies during the seizure.

4. REMOTE LOCAL ACTION AND USES.

Ether is excreted like chloroform, and to a certain extent

increases all the secretions, but is not employed with this end in view. It is believed by some to diminish the liability to gall stones, or actually to dissolve concretions already formed.

<u>NITROUS OXIDE.</u> $N_2O.$ (*Not Officinal.*)

Nitrous Oxide Gas, Protoxide of Nitrogen, "Laughing Gas."

Source.—Made by heating Nitrate of Ammonia. $NH_4NO_3 = N_2O + 2H_2O.$

Characters.—A colourless inodorous gas. It is provided for use condensed into a liquid, in strong iron bottles, whence it is allowed to escape into a caoutchouc bag.

ACTION AND USES.

1. ACTION ON THE BLOOD AND ITS USES.

Nitrous oxide gas, administered from an inhaler, rapidly enters the circulation; is absorbed by the plasma; converts the arterial into venous blood, in the course of about sixty seconds; and thus produces **partial asphyxia**. It does so apparently by diminishing the amount of oxygen in combination with the red corpuscles, without itself uniting with the hæmoglobin, like CO and NO; in this respect it is an "indifferent" gas, like N and H, simply taking the place of the oxygen, if this be completely excluded at the same time, and exerting of itself no poisonous action upon the corpuscles. It must, therefore, be given pure, *i.e.* without any admixture of air. The effect of the incipient asphyxia, and the use to which it may be turned, will be described in the next section.

2. SPECIFIC ACTION AND USES.

Nitrous oxide gas not only renders the blood venous, but simultaneously enters the nervous centres, upon which it acts, first as a stimulant, and speedily as an **anæsthetic**. Thus the gas produces a series of phenomena which can be resolved into the parallel effects of venosity of the blood or asphyxia, and a specific influence on the nerve cells of the convolutions. After a few seconds' excitement, the subject for anæsthesia by nitrous oxide begins to breathe laboriously; the mind becomes rapidly obscured; and by the end of about sixty seconds consciousness is lost, the face may be livid and bloated in appearance, respiration becomes stertorous, muscular twitchings occur, the pulse fails at the wrist, and the whole appearance is alarming to a novice. If the inhalation be now interrupted, perfect recovery of consciousness and of natural breathing occurs in thirty to sixty seconds, with disappearance of all the urgent symptoms. It is clear that asphyxia is carried into the second stage—that of respiratory excitement, but not beyond, neither the movements of the chest nor the action of the heart being arrested. But even if these untoward results should occur, resuscitation is easy by means of artificial respiration; it is said even after five minutes in the case of rabbits. No attempt to carry the asphyxia beyond the second stage is permissible in man.

Nitrous oxide gas is extensively used to produce anæsthesia during operations lasting but one minute or less, and especially by dental surgeons during the extraction of teeth, destruction of the nerve, etc. It must always be given pure, by the arrangement above described in the hands of a skilled anæsthetist. The moment for operating is best indicated by stertorous breathing and twitching of the muscles. Persons with diseased vessels, such as the subjects of chronic Bright's disease, ought not to take this anæsthetic, which produces (like all asphyxiating agents) a great and sudden rise of the arterial pressure, likely to cause rupture within the brain.

DICHLORIDE OF ETHIDENE. *(Not Officinal.)*

Source.—Obtained in the manufacture of chloral.

Characters.—A colourless volatile liquid, with the odour and taste of chloroform. Specific gravity, 1·20. Readily soluble in ether, chloroform, and alcohol; with difficulty in water.

ACTION AND USES.

Dichloride of ethidene is a general anæsthetic, supposed to occupy a position somewhat between ether and chloroform, but depressing the heart more than the latter. It is a very safe anæsthetic in some animals, but, like all its allies, occasionally causes death in man. About 4 fl.dr. in the form of vapour are required for an adult.

ÆTHYL BROMIDUM. Bromide of Ethyl.

Hydrobromic Ether. C_2H_5Br. *(Not Officinal.)*

Source.—Made by adding Bromine to a mixture of Phosphorus and Absolute Alcohol, and distilling. $5C_2H_6O + PBr_5$ (bromide of phosphorus) $= 5C_2H_5Br + H_3PO_4 + H_2O$.

Characters.—A colourless liquid with a powerful fragrant odour, and a hot sweetish taste. Very volatile; specific gravity,

1·42. Non-inflammable. Readily decomposes, yielding bromine. Freely soluble in alcohol and ether; very sparingly soluble in water.

Dose.—10 to 60 min.

ACTION AND USES.

Bromide of ethyl acts as an **anæsthetic** like chloroform and ether. For a time it was used in America and England, especially in short painful operations, and in ophthalmic practice, as its action is rapid and evanescent, and sickness rare. More than one death during or after its administration must account for its sudden loss of popularity. It has also been given by the stomach as an antispasmodic, especially in convulsions.

BICHLORIDE OF METHYLENE. CHLORIDE OF MONO-CHLOR-METHYL. $CH_2Cl_2(CH_2Cl.Cl)$.
(*Not Officinal.*)

Source.—Obtained from Chloroform by the action of nascent Hydrogen, one atom of which replaces one atom of chlorine in the Chloride of Dichlor-methyl (chloroform), $CHCl_2.Cl$.

Characters.—A colourless volatile liquid, with an odour like chloroform. Specific gravity, 1·344. Soluble in water, ether, and alcohol.

ACTION AND USES.

Bichloride of methylene acts as a **general anæsthetic** very much like chloroform. It is said, however, to depress the heart even more than this substance; and it is now very seldom used for general surgery, but is the anæsthetic most frequently employed in ovariotomy.

ETHYLATE OF SODIUM, SOLUTION OF.
(*Not Officinal.*)

Source.—Made by dissolving metallic Sodium in Absolute Alcohol.

Characters.—A brownish syrupy liquid.

ACTION AND USES.

Sodium ethylate is a powerful caustic, used to destroy small accessible tumours, such as nævi.

SPIRITUS ÆTHERIS NITROSI. SPIRIT OF NITROUS ETHER.

"Sweet Spirit of Nitre." A spiritous solution, containing nitrous ether, ethyl nitrite, $C_2H_5NO_2$.

Source.—Made by distilling a mixture of Rectified Spirit, Nitric Acid, Sulphuric Acid, and Copper; and dissolving the Distillate in Spirit. $C_2H_6O + HNO_3 + H_2SO_4 + Cu = C_2H_5NO_2 + CuSO_4 + 2H_2O$. The equation represents the formation of ethyl nitrite, but the drug also contains acetic ether, aldehyde, and acetic acid dissolved in spirit.

Characters.—Transparent and nearly colourless, with a slight tinge of yellow, mobile, of an apple-like odour, and a sweetish cooling sharp taste; slightly acid; inflammable. Sp. gr., ·845.

Incompatibles.—Iodide of potassium, sulphate of iron, tincture of guaiacum, gallic and tannic acids. Emulsions are curdled by its addition.

Impurity.—Excess of acid; effervescing much with $NaHCO_3$.

Dose.—$\frac{1}{2}$ to 2 fl.dr.

ACTION AND USES.

1. IMMEDIATE LOCAL ACTION AND USES.

In the stomach spirit of nitrous ether is a **diffusible stimulant** and carminative, doubtless from the amount of alcohol which it contains.

2. ACTION ON THE BLOOD.

The nitrite of ethyl appears to produce the same effect on the red corpuscles as other nitrites, especially diminishing oxygenation. *See* AMYL NITRIS.

3. SPECIFIC ACTION AND USES.

Although anæsthetic to a degree, sweet spirit of nitre chiefly acts upon the circulation like amyl nitrite. It relaxes the peripheral vessels, and accelerates the heart, but much less quickly, less completely, and more persistently than the amyl compounds. Thus it lowers **arterial tension**, and causes the phenomena described at page 163, only in a much less degree. By relaxing the renal vessels it is diuretic, the water alone being increased; by dilating the cutaneous vessels, as well as by perspiration, it increases the loss of heat from the skin. Nitrous ether is chiefly used as an **antipyretic** in febrile affections, where it diminishes the heat production by acting on the blood, and increases the loss of heat through the skin and kidneys. As a **diuretic** it is useful when a free watery flow is desired to wash out the tubules and passages, and relax spasm in the renal vessels, as in some cases of Bright's disease with increased arterial tension. Probably for the same reason it fails as a diuretic in cardiac dropsy, where the veins demand relief, and the arterial pressure is already too low. Being a dilator of the renal vessels, it must not be used in acute inflammatory states of the kidney. Spirit of nitrous ether may also relieve angina pectoris, and cardiac pain dependent on a failing and dilating heart in chronic Bright's disease. Like other nitrites, it has benefited some cases of dysmenorrhœa and of asthma.

4. REMOTE LOCAL ACTION.

This compound or its constituents are chiefly excreted by the kidneys and lungs. Its diuretic influence has just been described.

ÆTHER ACETICUS. ACETIC ETHER.

$$C_2H_5.C_2H_3O_2.$$

Source.—Made by (1) distilling Rectified Spirit with Acetate of Soda and Sulphuric Acid; and (2) separating the ethereal liquid by means of Chloride of Calcium. (1) $NaC_2H_3O_2 + H_2SO_4 + C_2H_6O = C_2H_5,C_2H_3O_2 + NaHSO_4 + H_2O.$

Characters.—A colourless liquid, with an agreeable ethereal, somewhat acetous odour, and refreshing taste. Specific gravity, 0·91. Neutral. Soluble freely in rectified spirit and ether, and in about 12 parts of water.

Dose.—20 to 60 min.

ACTION AND USES.

Acetic ether is a **stimulant** and **antispasmodic**, much like ether itself, but forms more agreeable combinations with other **carminatives** on account of its pleasant odour and taste.

CHLORAL HYDRAS. HYDRATE OF CHLORAL.

$$C_2HCl_3O.H_2O.$$

Source.—Made from Chloral by the addition of Water. Chloral (C_2HCl_3O) is itself made by the action of dried Chlorine upon anhydrous Alcohol, and purifying. $C_2H_6O + Cl_8 = C_2HCl_3O + 5HCl.$

Characters. — Colourless crystals, or white crystalline masses, with a peculiar pungent odour, and a pungent, rather bitter taste. Readily fused by gentle heat, recrystallising on cooling to 120°. Solubility, very freely in distilled water, rectified spirit, and ether; 1 in 4 of chloroform.

Incompatibles.—All alkalies.

Impurities.—Hydrochloric acid, detected by test-paper; oily substances, colouring sulphuric acid when dissolved in chloroform.

Dose.—5 to 30 gr.

Preparation.

Syrupus Chloral.—10 gr. in 1 fl.dr. *Dose*, ½ to 2 fl. dr.

ACTION AND USES.

1. IMMEDIATE LOCAL ACTION AND USES.

Externally.—Applied in weak solution (5 gr. to the ounce of water), chloral hydrate is antiseptic; concentrated solutions are irritant, causing vesication. The drug is but little used externally.

Internally.—In the mouth and stomach it is also irritant unless freely diluted, a point to be observed. It has no specially sedative effect on the stomach or bowels like opium, and therefore causes neither dyspepsia nor constipation.

2. ACTION ON THE BLOOD.

Chloral enters the blood as such, and probably leaves it for the tissues without decomposition, although Liebreich, who introduced it into the materia medica, contends that it is broken up into chloroform and formic acid in the presence of the sodium salts of the plasma. $C_2HCl_3O + NaHO = NaCHO_2 + CHCl_3$. The blood undergoes no appreciable change.

3. SPECIFIC ACTION AND USES.

The action of chloral upon the system so nearly resembles that of chloroform, and the chemical relations of the two substances are so close, that Liebreich's theory is at first sight extremely plausible. Chloral chiefly affects the nervous system, although one of the principal dangers connected with its use depends on its direct action on the heart. Given in moderate doses (20 to 30 gr.), hydrate of chloral, after a very brief period of excitement, quickly induces drowsiness, followed by several hours' sound sleep, natural in its characters and refreshing in its effect; as a rule, without consequent confusion, headache, or drowsiness in healthy individuals. Larger doses produce deeper and more prolonged sleep, and an appearance of narcosis, the subject being difficult to rouse even by sharp stimulation. Thus far chloral manifestly acts upon the convolutions, either directly or through the cerebral circulation, or both; and is a **pure and** powerful hypnotic. The larger doses, however, enable us to appreciate its action, like that of chloroform, on the lower nervous centres. The spinal centres are depressed, whence diminished reflex excitability and relaxation of the muscles. The three great medullary centres are decidedly depressed: respiration becomes slow, irregular, and shallow; the heart is weakened (but chiefly in another manner, as we shall presently find); and the vaso-motor centre is lowered in activity, so that the vessels dilate generally. The peripheral sensory nerves are not specially affected. Neither are the motor nerves, or the muscles themselves, directly depressed.

Upon these several effects of chloral depend at once its value medicinally, and the drawbacks or even dangers which occasionally attend its employment. It is the most rapid, and probably the most powerful, whilst the most pure, of all the hypnotics, opium not excepted. It is therefore extensively used to produce sleep and soothe the cerebral hemispheres, in conditions of excitement, in insomnia from over-work, distress, maniacal excitement, or despondency, and in the early stages of fevers or febrile diseases, whilst the heart is still strong. It is especially valuable in delirium tremens. In the sleeplessness which attends or is caused by peripheral pain, chloral fails, for an obvious reason; or if sleep be secured by a powerful dose, the patient wakes to suffering as before. To relieve the severe pain of neuralgia it is totally unfitted.

Chloral has also been given in the delirium of the more advanced stages of fevers; to relieve the distress, dyspnoea, and insomnia of cardiac and renal disease; and in the cough, spasm, and breathlessness attending phthisis, bronchitis, and other respiratory affections. The dangers of the drug in these conditions have been shown by the fatal results which have followed its employment; and the cause of them is obvious. Besides its depressing effect on the medulla, chloral in full doses acts as an intrinsic cardiac poison, slowing and enfeebling the heart by diminishing the irritability of its ganglia, and finally arresting it in ventricular diastole. At the same time the blood pressure falls by peripheral paralysis of the vessel walls, as well as from the interference with the vaso-motor centre, the heart, and the respiration; so that altogether the circulation tends to become arrested. Thus the relief to be obtained from chloral in the delirium of fever where the heart is threatening to fail, and in organic disease of the heart, lungs, or kidneys, is but temporary, and is purchased at serious cost; for this purpose the drug is not to be recommended.

The action of chloral in reducing the excitability of the grey matter of the cord, and higher motor ganglia, has suggested its use in tetanus, strychnia poisoning, puerperal convulsions, hydrophobia, sea-sickness, and whooping-cough. It has also been given in some cases of chorea, but here really as a hypnotic.

The exact effect of chloral on metabolism is unknown. It reduces temperature, chiefly by increased loss of heat from the dilated peripheral vessels, but also by diminishing the production in the weakened muscles, etc. It must not, however, be given as an antipyretic in high fever, unless at the commencement, in strong subjects, on account of its depressant action on the heart. It has been highly recommended in cholera.

4. REMOTE LOCAL ACTION.

Chloral is excreted by the kidneys partly unchanged, but chiefly as urochloralic acid producing slight diuresis. Probably part escapes by the skin also, as a variety of eruptions may attend its prolonged use.

5. ADVANTAGES AND DISADVANTAGES OF CHLORAL; CAUTIONS; CONTRA-INDICATIONS.

It will be well to state here succinctly the advantages and disadvantages of chloral as compared with morphia (opium). Chloral has the following *advantages*: It acts quickly as a hypnotic—even more quickly than morphia subcutaneously, and more certainly even when morphia has failed. After-effects, such as headache, depression, and sickness are less common from chloral. It does not derange the stomach, if freely diluted; nor cause constipation, even when given for a long time. It may be more safely given, in proper doses, to children.

On the other hand, chloral has these *disadvantages*: It does not relieve pain, and is thus greatly inferior to opium in most cases as a hypnotic, and useless as an anodyne. It does not, like opium, prevent or relieve distress, reflex dyspnoea, and cough due to heart and lung disease. Chloral causes excitement instead of quiet, in many cases of mania, hysteria, and confirmed alcoholism.

Chloral must be given in relatively small doses to children and delicate persons; and very rarely, as we have seen, to the subjects of organic disease of the heart, lungs, and kidneys, or patients suffering from gout. If it excite instead of soothing the insane or the confirmed drunkard, it should not be persevered with ; nor if it increases instead of relieving sleeplessness in certain individuals, as it does occasionally, apparently from idiosyncrasy. Lastly, chloral must be prescribed with great hesitation to persons who suffer from constitutional debility of the nervous system, expressing itself in hysteria, despondency, excitability, and innumerable other forms. Such subjects very readily acquire the " chloral habit," that is, they consume on their own account regular and ever increasing quantities of chloral, until the nervous system and general nutrition fail, the mind is demoralised, and the victims ultimately perish like the drunkard and opium eater.

BUTYL-CHLORAL HYDRAS. Hydrate of
Butyl-Chloral. Croton-Chloral. $C_4H_3Cl_3O.$
(*Not Officinal.*)

Source.—Made by passing Chlorine gas through Acetic Aldehyde, which is converted first into Crotonic Aldehyde, and then into Croton Chloral.

Characters.—Small brilliant tabular crystals, with a pungent odour, much like that of chloral hydrate, and an acrid nauseous taste. Solubility, 1 in 100 of water, freely in spirit, 1 in 4 of glycerine.

Incompatibles.—As of chloral hydrate.

Dose.—2 to 15 gr.

ACTION AND USES.

In every important respect the action of butyl-chloral is nearly allied to that of chloral hydrate, and it will therefore suffice to indicate the points wherein the two drugs differ.

Butyl-chloral as a hypnotic is less rapid, less certain, and less powerful than the other, which is generally to be preferred for this purpose. It is believed that butyl-chloral is less depressant to the heart, and therefore that it may be given in insomnia with cardiac weakness where chloral hydrate would be inadmissible. We must accept this recommendation with great caution. The most important effect of butyl-chloral, peculiar to itself, is **anæsthesia** of the region of the trigeminus, that is, of the face and part of the scalp, preceding the hypnotism. But even this action has been disputed by good authorities. The drug relieves some cases of *tic-douloureux* and facial

L—8

neuralgia very quickly; in some cases it fails. It has been given in other forms of pain in the face, such as toothache (locally); in neuralgia of the limbs; and in painful menstruation.

AMYL NITRIS. Nitrite of Amyl. $C_5H_{11}.NO_2$.

Source.—Produced by heating Nitric Acid with Amylic Alcohol, distilling, and purifying the product.

Characters.—An ethereal liquid, of a yellowish colour, and peculiar pine-apple odour. Sp. gr., ·877· Insoluble in water; soluble in rectified spirit, ether, and chloroform.

Impurities.—Nitric and hydrocyanic acids.

Dose.—2 to 5 min., used with caution as inhalation from a crushed capsule; or ½ to 1 min. internally, dissolved in rectified spirit.

ACTION AND USES.

1. IMMEDIATE LOCAL ACTION AND USES.

Applied *directly* to peripheral nerves and muscles, nitrite of amyl depresses or paralyses them. It is never so employed in man. *Internally*, the drug is seldom given by the mouth, except in cholera.

2. ACTION ON THE BLOOD.

Nitrite of amyl is almost invariably administered by inhalation, a few drops being kept ready for use in a glass capsule (enveloped in cotton wool), which may be broken between the fingers and thumb when required. The vapour instantly enters the circulation through the lungs, converts a certain amount of hæmoglobin into methæmoglobin, and thus interferes with the oxygenating function of the red corpuscles; the amount of oxygen absorbed (in animals) being quickly lowered, as well as the excretion of carbonic acid. The blood of animals killed by nitrite of amyl is of a chocolate colour; but the effect of an ordinary inhalation in man is very transitory.

3. SPECIFIC ACTION AND USES.

Nitrite of amyl almost instantaneously reaches the tissues, and produces striking phenomena. Two to five drops, inhaled as directed, immediately produce a sense of fulness and throbbing in the head; flushing of the face, neck, and trunk; tingling over the surface generally; dilatation of the pupils, and disturbance of vision; giddiness and unsteady gait;

increased frequency and force, that is, palpitation, of the heart; visible pulsation of the carotids; restlessness and anxiety of mind. These symptoms quickly disappear, leaving possibly slight headache. Larger doses aggravate all the phenomena, but never produce unconsciousness; mental confusion, intense bodily depression, coldness of the extremities, and sweats being the result, followed by severe headache which may last for hours. Very rarely convulsions occur in man as in some of the lower animals.

The specific action of nitrite of amyl proves on analysis to be almost confined to the circulatory system, the other parts being chiefly involved secondarily. Two distinct effects are produced on the circulation; the heart is greatly accelerated, with but little, if any, increase of its force; the peripheral vessels are dilated by relaxation of their muscular coat. Some authorities hold that the cardiac acceleration is due to depression of the cardiac centre, others to depression of the vagus in the heart; some refer the vascular relaxation to the action of the nitrite on the vasor centre in the medulla, others to its action on the vaso-motor nerves and muscular walls. Be this as it may, the fact remains that the blood pressure falls to a remarkable degree, that is, the resistance to the discharge of the left ventricle is correspondingly diminished; whilst this discharge is accomplished much more frequently within a given time. In other words, the left ventricle, under the influence of nitrite of amyl, has at once less work to accomplish, and more force wherewith to accomplish it; that is, is greatly relieved. These considerations led Dr. Lauder Brunton to the employment of the drug in those cases of the complex class of disease known as angina pectoris, in which agonising pain in the breast and neighbourhood is due to distension of the left ventricle, from its inability to empty itself against the pressure in the aorta, and in which fatal paralysis of the heart, or even rupture of its walls, is the result of the unequal effort. Clinical experience has fully confirmed the value of amyl nitrite, in cases where spasm of the arteries is damming the blood back upon the ventricle, the channels being instantly opened and the ventricle rapidly emptied by the double effect of the drug. The pain of the aneurism of the aorta, and of other forms of cardiac disease and disorder, can often be relieved by amyl, but caution must be exercised in the first trial. Threatening death from cardiac paralysis in chloroform anæsthesia, and sea-sickness in which the blood pressure is greatly disturbed, are sometimes successfully treated with amyl. Some cases of epilepsy, accompanied by spasm of the cerebral vessels and facial pallor, and of megrim or sick-headache, due to similar spasm in the trigeminal area, are also benefited by this drug.

The reflex irritability of the cord is reduced (in animals) by nitrite of amyl, which has therefore been proposed as a remedy in poisoning by strychnia. Neither the peripheral nerves nor the muscles are affected when it is given through the blood. Respiration is disturbed, apparently by the alteration of the hæmoglobin and circulation, not through the nervous system. The nitrite sometimes affords immediate relief in asthma, but the dyspnœa may as quickly return. The body temperature falls, from obvious causes.

4. REMOTE LOCAL ACTION.

Nitrite of amyl probably escapes from the body by the urine, which is decidedly increased in amount, and may contain sugar. Both of these effects are probably results of local disturbance of the pressure in the kidneys and liver respectively.

NITRO-GLYCERINUM.   NITROGLYCERINE.
$C_3H_5N_3O_9$.   (*Not Officinal.*)

Source.—Made by dropping Glycerine into a mixture of Sulphuric and Nitric Acids, washing, and evaporating to a proper density.

Characters.—An oily liquid, colourless, odourless, with a sweet pungent taste. Specific gravity, 1·60. Slightly soluble in water; freely in fats, oil, alcohol, and ether. Highly explosive, and, when mixed with infusorial earth, constitutes dynamite. Never used in the pure state.

Preparation.

Liquor Nitroglycerine.—1 gr. in 100 min. of rectified spirit. *Dose*, ½ to 2 min.; in water, or as a chocolate tablet.

ACTION AND USES.

This powerful substance closely resembles in its action the nitrites of amyl and soda; its activity being apparently due to nitrous acid formed by its decomposition within the body. It is used for the same class of cases, such as angina pectoris, chronic heart disease, sea-sickness, and asthma and other spasmodic disorders.

ÆTHYL IODIDUM. IODIDE OF ETHYL. HYDRIODIC ETHER. C_2H_5I. (*Not Officinal.*)

Source.—Made by adding iodine to a mixture of Alcohol and Phosphorus, distilling, and purifying. $5C_2H_5HO + PI_5$ (iodide of phosphorus) $= 5C_2H_5I + H_3PO_4 + H_2O$.

Characters.—A colourless volatile liquid, with a peculiar powerful odour, and a pungent taste. Sp. gr., 1·92. Non-inflammable. Decomposed by light, yielding iodine. Soluble in alcohol and ether; very sparingly soluble in water.

Dose.—5 to 20 min. Inhaled from a broken capsule.

ACTION AND USES.

Iodide of ethyl acts chiefly by virtue of the iodine element, and very slightly, if at all, as an anæsthetic ethyl compound. It introduces iodine very rapidly into the system, and has been chiefly used to stimulate the respiratory passages, and thus to act as an **antispasmodic** in asthma attended by scanty or tough secretion. In some instances it gives instant relief.

ACIDUM HYDROCYANICUM DILUTUM.

DILUTED HYDROCYANIC ACID.

"Prussic Acid." Hydrocyanic acid, HCN, dissolved in water, and constituting 2 per cent. by weight of the solution.

Source.—Made by distilling solutions of Ferrocyanide of Potassium and Sulphuric Acid. $2K_4Fe(CN)_6 + 6H_2SO_4 = 6HCN + Fe_2K_2(CN)_6 + 6KHSO_4$.

Characters.—A colourless liquid, with a peculiar penetrating odour. Sp. gr., 0·997. Faintly acid.

Incompatibles.—Salts of silver, copper, iron, red oxide of mercury, and sulphides.

Impurities.—Sulphuric and hydrochloric acids.

Dose.—2 to 8 min.

Preparation.

Vapor Acidi Hydrocyanici.—10 min. in 1 fl.dr. cold water.

Hydrocyanic acid is also contained in Aqua Laurocerasi. See also *Amygdala Amara*, page 239.

ACTION AND USES.

1. IMMEDIATE LOCAL ACTION AND USES.

Externally.—Applied for a time to the skin, dilute hydrocyanic acid causes numbness, directly depressing the sensory nerves. It is used, largely diluted, to relieve itching, but must not be employed where the surface is raw from scratching, as it is readily absorbed by wounds.

Internally, it produces a peculiar mixed sensation on the mouth and throat, and acts as a sedative to the nerves of the stomach. It is in common use to relieve pain and arrest vomiting in painful dyspepsia, ulcer of the stomach, cardialgia, and reflex or other nervous cases, *e.g.* in phthisis and pregnancy. The specific action of the drug on the medulla, to be presently described, doubtless assists its local effect upon the gastric nerves in producing these results.

2. ACTION ON THE BLOOD.

Hydrocyanic acid enters the blood very rapidly from all parts, especially the lungs, and produces an important change on the red corpuscles. If freely given, it converts the blood of the veins first into a bright arterial colour, and then into a deep black, the former change arresting the oxygenating function of the corpuscles, the latter destroying them. When studied in drawn blood, these effects are found to be due partly to *reduction* of the oxyhæmoglobin, the oxygen being replaced by cyanogen, forming cyano-hæmoglobin; and partly to *union* of the cyanogen with oxyhæmoglobin, making cyano-oxyhæmoglobin. Thus changed, blood does not give up oxygen to oxydisable bodies, *e.g.* the guaiacum reaction cannot be obtained. These effects of hydrocyanic acid on drawn blood must not be too readily applied to the circulating fluid within the body, where its action in medicinal doses is chiefly local and specific.

3. SPECIFIC ACTION AND USES.

Hydrocyanic acid rapidly enters the tissues, and acts chiefly upon the nervous structures. Considerable doses cause giddiness, faintness, nausea, a constricted feeling in the chest, headache, mental confusion, disturbed breathing, slowing of the pulse, and muscular debility. Larger doses aggravate these symptoms, and produce great dyspnoea and other signs of asphyxia; whilst in still larger quantity it is familiar as one of the most swift and deadly of poisons. Analysis proves that this drug, whilst depressing all nervous tissues, acts first and chiefly upon the *respiratory centre*, which is briefly excited and then depressed, leading to weak respirations with long pauses, dyspnœa, convulsions, and finally death by asphyxia. Simultaneously, the afferent branches of the *respiratory nerves* are depressed, especially if the acid be inhaled; and reflex respiratory acts are arrested. The *vaso-motor centre* is temporarily stimulated, and the blood pressure rises, but it falls again suddenly and greatly. The *cardiac centre* is the most resistant of the three, but it also is depressed, so that the action of the heart becomes less frequent and powerful. Although at the same time the nervo-muscular structures of the heart are depressed, the heart continues to beat in animals poisoned with prussic acid, after the respiration and other functions have-ceased. The *convolutions* are depressed, causing stupor, ending in unconsciousness; but this effect may be secondary to the disturbance of respiration. The cord is also lowered in activity. The peripheral *sensory nerves* are but little affected by the internal use of the drug, compared with its effect upon them locally. The *motor nerves* and *muscles* are depressed by repeated small doses of dilute hydrocyanic acid, the influence extending downwards.

The chief specific use of this drug is to allay dry, useless cough, by its action on the respiratory centre and the afferent nerves, in phthisis, pertussis, and asthma. In phthisis it also checks the tendency to cough and vomit induced by food. As a cardiac sedative, it is employed in the palpitation, pain, and distress brought on by dyspepsia, where again it fulfils a double indication. Its general sedative effect on the nervous system has suggested its use in epilepsy, chorea, hysteria, and tetanus, but with very doubtful benefit.

4. REMOTE LOCAL ACTION.

The mode of excretion of hydrocyanic acid is still obscure. Probably it escapes in part, as it enters in part, by the lungs; and some of it is supposed to be thrown out as formate of ammonia.

TRIMETHYLAMIN. PROPYLAMIN. (*Not Officinal.*)

An ammonia compound, in which the three atoms of hydrogen are replaced by methyl, $NC_3H_9 = N(CH_3)_3$, dissolved in water.

Source.—Obtained from herring brine by distillation. It is also contained in cod-liver oil and in various plants.

Characters.—Trimethylamin contains 10 to 20 per cent. of the ammonia compound; is colourless, with a very disagreeable smell and taste; alkaline; miscible with water. Propylamin is another name for impure trimethylamin.

Dose.—20 to 60 min.

ACTION AND USES.

Externally, trimethylamin is a local irritant. *Internally*, it lowers the frequency of the heart, the blood-pressure, and the temperature. It was for a time believed to be specially valuable in acute rheumatism, but is now very seldom used.

ACIDUM CARBOLICUM. CARBOLIC ACID, PHENIC ACID, PHENOL, PHENYLIC ALCOHOL. C_6H_5HO.

An acid obtained from coal-tar oil by fractional distillation and purification.

Characters. — Colourless, acicular crystals; hygroscopic; with a tarry odour and burning taste. Becomes and remains fluid on addition of 6 per cent. of water. Solubility, 1 in 15 of water, 1 in 1¾ olive oil, 3½ in 1 of glycerine. It does not redden blue litmus paper.

Dose.—1 to 3 gr.

Preparations.

1. **Glycerinum Acidi Carbolici.**—1 to 4.

2. **Suppositoria Acidi Carbolici cum Sapone.**—1 gr. in each.

Sodæ Sulphocarbolas.—Sulphocarbolate of Soda. $NaC_5H_6SO_4H_2O$. (*Not Officinal.*)

Source.—Sulphocarbolic acid is formed by direct union of Carbolic and Sulphuric Acids. The Sulphocarbolate of Soda by neutralising Sulphocarbolic acid with Carbonate of Soda.

Characters.—Whitish odourless lumps of minute colourless rhombic prisms. Soluble in water.

Dose.—20 to 60 gr.

ACTION AND USES.

1. IMMEDIATE LOCAL ACTION AND USES.

Externally.—The principal action and uses of carbolic acid in disease depend upon its influence on fermentation and decomposition, which are intimately associated with many pathological processes. When this influence is studied apart from the body, we find that most *organised ferments* (fungi, bacteria, infusoria) are readily deprived of their characteristic powers by solutions of carbolic acid; whilst *chemical ferments*, such as pepsin and ptyalin, are much less readily affected. At the same time the products of decomposition, which are almost invariably foul-smelling, are *deodorised* by the phenol. The exact *modus operandi* in all these cases is still unknown, as are also the nature of the processes, and the relation of organisms to them. Be the explanation what it may, the power of carbolic acid, or of any substance which can thus arrest molecular processes universally at work in physiology and pathology, must be said to be enormous, both in itself and in its effects.

Carbolic acid is extensively employed in the **antiseptic** method of the treatment of wounds, associated with the name of its introducer, Sir Joseph Lister. A 5 per cent. solution in water serves as a spray to cleanse instruments, and to wash the skin of the part before operation. A $2\frac{1}{2}$ per cent. watery solution is used to purify sponges and the hands of the operator, and as a lotion. Dissolved in olive oil (1 to 10, 1 to 20, 1 to 50, or still weaker), or 1 part of carbolic acid with 7 parts of castor oil and 8 of almond oil, it is used for lubricating catheters, or as a special dressing. Carbolic acid gauze consists of unbleached cotton gauze medicated with half its weight of a mixture of carbolic acid (1), resin (4), and paraffin (4).

Coming to its physiological action proper on the human tissues, we find that carbolic acid is a local irritant to the skin, causing burning pain, anæsthesia, and finally a caustic effect with formation of a white hard eschar. It may therefore be applied to poisoned wounds and foul ulcers; and in dilute solutions is a stimulating as well as a disinfecting wash to wounds and discharging mucous surfaces or cavities, in the form of a lotion, injection, or gargle. It also relieves some forms of itching and inflammatory skin diseases, and is used with success in ringworm, where it destroys the vegetable organisms.

Apart from the body, carbolic acid is extensively used as **a general disinfectant.**

Internally.—In the form of vapour, carbolic acid is stimulant and disinfectant, and is used in ulceration of the throat and lungs (phthisis, dilated bronchi, gangrene, etc.), much importance having lately been attached to it in the so-called "antiseptic" treatment of phthisis. In the stomach and bowels it is a powerful irritant poison in large doses; in moderate quantity it arrests fermentative changes in the gastric contents in cases of dilatation of the viscus. Two other points may be noted in this connection; first, that carbolic acid unites with sulphates to form sulphocarbolates, which suggests the use of soluble sulphates as antidotes in poisoning by the drug; and secondly, that phenol is a natural product of the intestine or its contents.

2. ACTION ON THE BLOOD.

Carbolic acid is rapidly absorbed from the unbroken skin, mucosæ, wounds, subcutaneous tissues, respiratory passages, and stomach; and for a considerable time can be found in the blood unchanged. Here it steadily disappears, by conversion into compounds from which it may be again derived; uniting, for example, with sulphates, as already described. The blood is dark and slow to coagulate after poisoning by the drug.

3. SPECIFIC ACTION AND USES.

The action of carbolic acid on the organs is of little interest to the therapeutist. It is found in them chiefly as phenol-yielding compounds; and its effects in man are chiefly those of an irritant poison. The heart first falls and then rises in frequency, from disturbance of the cardiac centre. The blood pressure rises at first, returns to the normal, and falls after a fatal dose. Dyspnœa ensues, also central in origin. Convulsions occur in the lower animals through the cord, then paralysis and collapse. The voluntary muscles are not affected by carbolic acid, but the pupil is contracted. Sensibility is not reduced by internal administration of the drug. The temperature falls slightly after medicinal doses, but may rise in cases of dangerous absorption from dressings. Carbolic acid has been given internally in fevers, it is said with good results, but is little used in this country for such a purpose. It may temporarily relieve diabetes.

4. REMOTE LOCAL ACTION.

Carbolic acid and its products rapidly leave the body, chiefly by the urine. But little of it can be recovered unchanged, for (1) part is lost in the system, being probably converted into oxalates and carbonates; (2) part appears as sulphocarbolic acid ($C_6H_5.H.SO_4$) in combination; (3) part is constituted by an obscure compound; and (4) the remainder appears to give rise to a peculiar olive-green, brown, or grey discoloration of the urine, which is familiar to surgeons. It is important to note that this change in the urine bears no definite relation to the amount of carbolic acid in the blood, or the danger of poisoning. Fainting and collapse are the principal symptoms of its excessive absorption from a wound or through the skin, with or without rise of temperature. Disappearance of the sulphates from the urine, easily ascertained by ordinary tests, is a sure indication of danger. Albuminuria is sometimes induced.

Carbolic acid also leaves the body by the saliva, which is increased, and it stimulates the flow of sweat although it is not found in it.

Resorcin. (*Not Officinal.*)—A derivative of carbolic acid by various processes.

Characters.—White tabular lustrous crystals, with a weak odour like carbolic acid, and a sweetish, somewhat pungent taste. Solubility, 1 in 2 of water; 1 in 20 of olive oil.

Dose.—5 to 30 gr. every two hours, or single doses of 60 gr. at long intervals.

ACTION AND USES.

Externally.—Resorcin is antiseptic and disinfectant, without being irritant, in ordinary solutions (2 to 10 per cent.) It has been used as a dressing for all kinds of sores and wounds.

Internally.—It passes rapidly through the system, and is excreted unchanged in the urine by the end of one hour. It causes diaphoresis, and reduces the temperature and pulse for a time in conditions of fever; but has no influence on the normal body heat. Excessive doses cause trembling, singing in the ears, deafness, and mental disturbance. It has been used as an antipyretic in fevers of every kind, and is said to be specially useful in ague; but the drug is still on its trial.

Chinolin. $C_9H_7N.$ (*Not Officinal.*)—A derivate of Cinchona Bark, whence its name. Now made synthetically, or by the action of glycerin on nitro-benzol and aniline, in the presence of a dehydrating agent.

Characters. — A colourless, oily-like, highly refracting liquid, with a peculiar odour. It forms salts with acids, of which the tartrate is used, occurring in lustrous crystals, soluble in water.

Dose.— Of chinolin, 3 to 10 min.; of the tartrate, 5 to 15 gr.

ACTION AND USES.

Chinolin is antiseptic and disinfectant externally. *Internally*, it is an antipyretic, like resorcin, and has been used for the same class of cases.

Kairin. (*Not Officinal.*)—A pale-buff powder, with a disagreeable bitter and somewhat aromatic taste. It is made from phenol.

Dose.—5 gr. every hour; or 15 to 20 gr. in single doses.

ACTION AND USES.

Kairin is **antiseptic, disinfectant,** and especially **antipyretic,** like the two preceding substances, to which it is closely allied.

Fuchsine—Magenta Dye. (*Not Officinal.*)—An aniline product occurring in brilliant beetle-coloured needles, forming an intense deep-red solution with water.

Dose.—$\frac{1}{2}$ to 4 gr.

ACTION AND USES.

Fuchsine passes through the blood and tissues, and colours the urine and fæces. It has been said to reduce the amount of albumen in some cases of Bright's disease.

Creasotum—Creasote.—A product of the distillation of Wood Tar.

Characters.—A liquid, colourless or with a yellowish tinge, a strong empyreumatic odour, and burning taste. Soluble, sparingly in water; freely in alcohol, ether, and glacial acetic acid. Sp. gr., 1·071. Coagulates albumen.

Impurity.—Carbolic acid; detected by becoming solidified by cooling.

Dose.—1 to 3 drops, with mucilage or bread crumb.

Composition.—Creasote is not a simple body, but a variable compound of *Guaiacol*, $C_7H_8O_2$, and *Kreasol*, $C_8H_{10}O_2$.

Preparations.

1. **Mistura Creasoti.**—Creasote, 16 min.; glaciaı acetic acid, 16 min.; spirit of juniper, 30 min.; syrup, 1 fl.oz.; water, 15 fl.oz. *Dose*, 1 to 2 fl.oz.

2. **Unguentum Creasoti.**—1 in 9.

3. **Vapor Creasoti.**—12 min. in 8 fl.oz. of boiling water.

ACTION AND USES.

The action of creasote is, practically speaking, the same as that of carbolic acid, to which the student is referred. Before the latter came into general use, creasote was not unfrequently employed for the same purposes internally to which carbolic acid is now put; but the uncertainty of its composition and strength, as a complex product, renders it inferior to phenol in this respect.

The Unguentum is employed in dry skin diseases. The Vapor is disinfectant and deodorant in phthisis, chronic bronchitis, gangrene, and other diseases of the lungs attended by foul discharges. A combination of creasote, iodine, and various volatile substances such as ether, chloroform, and spirit, has lately become popular as a constant inhalation in phthisis. The Mistura Creasoti is intended chiefly as a remedy in vomiting, especially when this is due to pyloric obstruction, dilatation of the stomach, and the development of fermentation; but it has also been recommended in the vomiting of pregnancy, hysteria, and sea-sickness.

A specific and remote local effect has lately been claimed for creasote, when given by the stomach, namely, as a disinfectant and deodorant in phthisis with antipyretic and healing properties.

IODOFORMUM. Iodoform. CHI_3. (*Not Officinal.*)

Source.—Made by the action of Iodine on a hot solution of Carbonate of Soda in Diluted Alcohol. $C_2H_6O + 8I + 3Na_2CO_3 = CHI_3 + NaCHO_2 + 5NaI + 2H_2O + 3CO_2$.

Characters.—Small lemon-coloured lustrous crystals, with a powerful and persistent saffron-like odour, and an unpleasant sweetish taste. Insoluble in water; soluble freely in fixed and volatile oils and ether. It contains more than 90 per cent. of iodine.

Dose.—$\frac{1}{2}$ to 3 gr. or more.

Preparations.

Unguentum Iodoformi (U.S.P.).—1 to 9 of benzoated lard. Iodoformum Præcipitatum.—An impalpable yellow powder. Iodoform Wool.—Absorbent Cotton Wool, containing 10 per cent. of iodoform.

ACTION AND USES.

1. IMMEDIATE LOCAL ACTION AND USES.

Iodoform is an antiseptic and disinfectant, but destroys organisms less readily than carbolic acid. It is a very powerful deodorant. Applied to the human tissues, it produces little or no irritation.

Iodoform is used to cleanse foul ulcers, especially of venereal origin; and may possibly have a special effect on strumous ulceration. It has also been extensively applied as an antiseptic dressing to healing wounds, the best forms being the wool and the ointment. Sometimes iodoform gauze has been employed. Iodoform bougies for insertion into the urethra and os uteri have not given satisfaction. A powder of iodoform diluted with quinine or bismuth is a valuable insufflation in ozœna and ulcers of the mouth and throat.

2. ACTION ON THE BLOOD, SPECIFIC, AND REMOTE LOCAL ACTION AND USES.

Iodoform is occasionally absorbed from wounds, causing sickness and fever, restlessness and delirium in some subjects, drowsiness and collapse in others. Iodine is possibly set free in the blood, appearing in the urine as iodide of sodium. Iodoform has been used in an endless variety of diseases internally, but unfortunately with no special benefit.

Petrolatum (U.S.P.) — PETROLEUM OINTMENT, VASELINE. (*Not Officinal.*)

Source.—Obtained from American petroleum by distilling off the lighter portions and purifying the residue.

Characters.—A yellowish semi-solid fat-like mass, transparent, odourless, tasteless, neutral. Insoluble in water; freely soluble in fixed and volatile oils.

ACTION AND USES.

Vaseline cannot become rancid or irritant to the skin, and, being readily miscible with many active substances, such as the phenol compounds and alkaloids, is indicated as a valuable basis for ointments instead of lard. Its chief disadvantage is the low point at which it melts, and its consequent tendency to spread through the dressings. It is now extensively used.

Part II.

THE ORGANIC MATERIA MEDICA.

GROUP I.
THE VEGETABLE KINGDOM.

RANUNCULACEÆ.

Aconiti Folia—ACONITE LEAVES.—The fresh leaves and flowering tops of Aconitum Napellus. Gathered when about one-third of the flowers are expanded, from plants cultivated in Britain.

Characters.—Leaves smooth, palmate, divided into five deeply cut wedge-shaped segments; exciting slowly, when chewed, a sensation of tingling. Flowers numerous, irregular, deep blue, in dense racemes.

Aconiti Radix—ACONITE ROOT.—The dried root of Aconitum Napellus. Imported from Germany, or cultivated in Britain. Collected in the winter or early spring before the leaves have appeared.

Characters.—Usually from one to three inches long, not thicker than the finger at the crown, tapering, blackish-brown, internally whitish. A minute portion, cautiously chewed, causes prolonged tingling and numbness.

Substance resembling Aconite Root: Armoracea. (*See* page 202.)

Composition.—The active constituent of aconite is *aconitia* or *aconitin*, $C_{30}H_{47}NO_7$, an amorphous or crystalline alkaloid, forming salts with acids. The names of *pseud-aconitin, napellin, nepallin, napalin, aconellin,* etc., have been given to other more or less identical active principles obtained from the same plant or its botanical allies. They are combined with a peculiar acid, *aconitic acid.*

Preparations.

A. *Of the Leaves :*

1. **Extractum Aconiti.**—A green extract. *Dose,* 1 to 2 gr.

B. *Of the Root :*

1. **Tinctura Aconiti.**—1 in 8 of spirit. *Dose*, 5 to 10 min.

2. **Linimentum Aconiti.**—1, in 1 of spirit, with $\frac{1}{20}$ camphor.

3. **Aconitia.**—An alkaloid obtained from aconite root. Made (1) by dissolving the alcoholic extract of the powdered root in water ; (2) Precipitating the impure aconitia by ammonia ; (3) Extracting the dried precipitate with ether, dissolving in diluted sulphuric acid, again precipitating with ammonia, and purifying.

Characters.—See *Composition.*   Not given internally.

Preparation.

Unguentum Aconitiæ.—1 in 60.

ACTION AND USES.

1. IMMEDIATE LOCAL ACTION AND USES.

Externally.—Applied to the skin, or an exposed mucous membrane, aconite affects the terminations of the sensory nerves, causing tingling, followed by numbness, and **lowering** the sensibility of touch and temperature. It is, therefore, used to relieve pain due to disorder of the peripheral nerves, especially certain forms of neuralgia, and acute and chronic rheumatism. The aconitia ointment must be employed with caution.

Internally.—Aconite and aconitia cause an intensely acrid sensation on the tongue, followed by persistent tingling and numbness. A sense of warmth, pain, and sickness follow its admission to the stomach in full doses.

2. ACTION ON THE BLOOD.

Aconitia enters the blood, and thence finds its way to the tissues.

3. SPECIFIC ACTION AND USES.

Medicinal doses of aconite, taken in close succession, reduce the frequency, force, and tension of the **pulse** ; flush and **moisten the skin** ; and **increase** the amount of **urine.** Larger doses cause a sense of illness and muscular weakness ; " creeping," " tingling," " numb " sensations generally, but especially on the lips, face, and extremities, ending in anæsthesia ; and disturbances of vision, hearing, and consciousness. On analysis, it is found that the heart is briefly accelerated, and then reduced in frequency through the nerves ; its force is then reduced, by direct action on the nervo-muscular structures; and finally the cardiac action becomes more frequent, irregular, and more and more feeble, tending to cease in diastole. The blood pressure falls continuously, partly from cardiac, partly from vaso-motor depression. Respiration is slowed and deepened, with spasmodic irregularity of rhythm, and finally is arrested after poisonous quantities. The skin is stimulated, perspiration becoming abundant. The kidneys are also stimulated, the fluids and solids of the urine being increased in amount. The **temperature falls** steadily. The muscular weakness appears to be primarily due to depression of the motor-nerve endings; but this condition extends to the cord. The brain itself is not directly affected, and even in cases of poisoning by aconite, consciousness is preserved almost to the end. The sensory nerves are probably paralysed from their periphery inwards by the internal, as by the external, administration of the drug.

Such being the specific action of aconite, its use is obviously indicated in the treatment of two conditions, namely, **fever and pain.** The cardio-vascular excitement, the dry skin, the high temperature, and the scanty secretions of fever, will all be relieved by this drug. For this purpose the tincture is given in small and closely repeated doses, say 1 minim in water every 15, 20, or 30 minutes, the effect being watched. Acute tonsillitis, bronchitis, pleurisy, and febrile conditions attending other local inflammations, have been treated with aconite, the effect being to control the urgent symptoms, relieve the distress of the patient, and even to cut short the disease. Some of the symptoms of scarlatina and measles may be similarly alleviated. The powerfully depressant action of aconite on the circulation altogether forbids its use as an antipyretic in heart disease, and suggests caution in its employment in all cases.

In neuralgia and other painful affections connected with the nerves and muscles, aconite may be given internally instead of being locally applied; facial neuralgia with spasm (*tic-douloureux*) particularly being relieved by it. In these cases, also, the tincture should be given in minim doses, repeated three or four times in an hour, and the effect watched.

4. REMOTE LOCAL ACTION AND USES.

Aconite is probably excreted by the kidneys, and, as we have already seen, increases the activity of their secretion. The stimulation of the sweat-glands and the occasional appearance of an eruption suggest that it also leaves the body by the skin.

M—8

Podophylli Radix—Podophyllum Root.—
The dried rhizome of Podophyllum peltatum. Imported from North America.

Characters.—In pieces of variable length, about two lines thick, mostly wrinkled longitudinally, dark reddish-brown externally, whitish within, breaking with a short fracture; accompanied with pale brown rootlets. Powder yellowish-grey, sweetish in odour, bitterish, subacrid and nauseous in taste.

Composition.—The active principle of the rhizome is the resin, which is really a compound of several resinous bodies.

Preparation.

1. **Podophylli Resina.**—Resin of Podophyllin.
   Source.—Made by extracting with spirit, and precipitating in acidulated water.
   Characters.—A pale greenish-brown, amorphous powder, soluble in rectified spirit and in ammonia.
   Dose.—$\frac{1}{4}$ to 1 gr.

ACTION AND USES.

Externally, podophyllin possesses no local action; but if applied to a wound, it enters the blood and exerts its specific effect as a purgative.

Internally, podophyllin causes a bitter acrid taste, salivation, irritation of the stomach, nausea, colic, and after ten or twelve hours a free watery motion. This purgative effect appears to be due to stimulation both of the muscular coat and of the glands of the intestine, as well as to increase of the biliary flow.

Podophyllin is used entirely as a purgative. One-grain doses are given to produce free evacuation of the bowels in severe constipation or portal congestion. A dose of $\frac{1}{6}$ to $\frac{1}{4}$ grain may be employed as an ingredient of habitual laxative pills. It is a useful cholagogue when mercurials are contra-indicated. Podophyllin must not be given alone, on account of its griping tendency, but combined with a carminative such as hyoscyamus, belladonna, or cannabis indica. The comparative slowness of its action must also be remembered.

MAGNOLIACEÆ.

Illicium Anisatum—Star Anise.—The oil distilled in China from the fruit forms part of the Oil of Aniseed of the Pharmacopœia. See *Oleum Anisi*. N.O. Umbelliferæ.

Actæa Racemosa. Cimicifuga. (*Not Officinal.*) — The rhizome and rootlets of Cimicifuga racemosa (Actæa racemosa) Black Snake-root or Black Cohosh. From the United States.

Characters.—Knotted heads, with numerous fine brittle rootlets; odour faint; taste somewhat bitter, astringent, and acrid, not unlike opium.
Composition.—Cimicifuga contains a *volatile oil*, two *resins*, *tannin.* The active principle is still uncertain.
Dose.—20 to 30 gr.

ACTION AND USES.

In moderate doses black snake-root is bitter; in larger doses it slows the heart and raises the blood pressure like digitalis; finally it is excreted in the urine, and increases the activity of the skin, kidneys, and generative organs.

Cimicifuga is used as a **stomachic**; in diseases of the heart; and in rheumatism, bronchitis, uterine disorders and spermatorrhœa, in which its remote stimulant action is occasionally valuable. It has been much lauded in chorea.

MENISPERMACEÆ.

Calumbæ Radix—CALUMBA ROOT.—The root, cut transversely and dried, of Jateorrhiza Calumba and Miersii. From the forests of eastern Africa, between Ibo and the Zambesi.

Characters.—Slices, flat, circular, or oval, about two inches in diameter, and from two to four lines thick, softer and thinner towards the centre, greyish-yellow, bitter. A decoction, when cold, is blackened by the solution of iodine.
Composition.—Calumba contains a non-nitrogenous, bitter principle, *calumbin*, $C_{21}H_{22}O_7$, crystallising in white needles; an alkaloid, *berberin*, $C_{20}H_{17}NO_4$; *calumbic acid*, $C_{21}H_{21}O_7$; 33 per cent. of *starch*; but no tannin.
Dose.—5 to 20 gr.

Preparations.

1. Extractum Calumbæ.—Aqueous. 8 in 1. *Dose*, 2 to 10 gr.

2. Infusum Calumbæ.—1 in 20 of cold water. *Dose*, 1 to 2 fl.oz.

3. Tinctura Calumbæ.—1 in 8. *Dose*, ½ to 2 fl.dr.

Calumba is also an ingredient of Mistura Ferri Aromatica.

ACTION AND USES.

Calumba is the first of the large and important group of bitter substances or **bitters**, which we meet with in the materia medica, and will therefore be fully discussed as the type of this class of remedies.   Under the head of the other bitters, such as quassia and gentian, fresh description of their action and uses will be unnecessary, and reference will simply be made to the present account.   So with the action and uses, *as bitters*, of the alkaloids (strychnia, quinia, etc.), and the aromatic bitters, including orange, lemon, cascarilla, etc.

1. IMMEDIATE LOCAL ACTION AND USES.

Externally.—Calumba and other bitters are antiseptic and disinfectant to a degree, arresting decomposition and fermentation.   They are not used for this purpose.

Internally.—Taken into the mouth, bitters, as their name implies, stimulate the nerves of taste, and thus induce several general reflex effects, of the first importance in digestion.   (1) The *saliva* is increased, and therewith its solvent and digestive influence on the food in the mouth, as well as its stimulant action on the gastric secretion; (2) The *vessels and glands of the stomach* are excited through the central nervous system, and the gastric secretion is thus increased in a second way; an effect which is heightened if the bitter be aromatic, and relish given by the pleasant flavour.

Reaching the stomach, calumba and other bitters stimulate digestion in a third way, by acting upon the gastric nerves and causing a sensation closely resembling hunger.   This rouses the appetite, and if food be taken within a few minutes, the other effects just described afford the means of digesting it. As in the mouth, the action of bitters in the stomach is greatly assisted by aromatics (essential oils) and alcohol (contained in tinctures).   Like these substances, bitters also stimulate the local circulation, and produce a remote effect on the heart and systemic vessels, raising the blood pressure, and thus acting as "general tonics."   They will also exert a certain controlling effect on any decomposition or fermentation which may be set up in the stomach.   When given in excess, or for a long time, bitters will manifestly, for every reason, tend to irritate the stomach and induce indigestion.

Calumba and bitters in general pass slowly along the intestines, moderating decomposition, and slightly stimulating peristalsis when they contain tannin, which many of them do. They are not cholagogue.

The *uses* of calumba and other bitters internally depend on the actions just described. They are of great value as stomachics, and much employed in rousing gastric digestion in atonic dyspepsia, where the appetite and the ability to digest have been diminished or lost, as in anæmia, convalescence from acute diseases, in persons exhausted by over-work, whether mental or bodily, and in the subjects of chronic constitutional diseases, such as phthisis and syphilis. In such cases, bitter infusions form the best vehicle for acid or alkaline stomachics, as the case may require, combined with an aromatic tincture, which renders the mixture much more agreeable and active. Their use must not be continued too long without intermission; they must not be given in too concentrated a form; and they must be employed with caution, or entirely avoided, in cases of dyspepsia attended by much pain, vomiting, mucous secretion, as well as in organic disease of the stomach. Calumba is one of the least irritant of all bitter stomachics.

The action of bitters on the bowels no doubt adds to their value in indigestion, as they remove flatulence and promote evacuation. Some forms of diarrhœa are relieved by calumba. Whether given by the mouth or as enema, bitter infusions are anthelmintic, preventing and destroying the thread-worm.

2. ACTION ON THE BLOOD, SPECIFIC ACTION, AND REMOTE LOCAL ACTION.

Whether bitters possess any *direct* action on the blood or tissues beyond those just described, is uncertain. The *indirect* effect on the system is manifestly great and of the first importance therapeutically, as they are the means of introducing into the blood an increased amount of nutrient material. In this way bitters are tonics, invigorating the body whilst they increase appetite; a system of treatment which is agreeable and striking to invalids and persons enfeebled by disease, over-work, or dyspepsia.

Pareiræ Radix—PAREIRA ROOT.—The dried root of Cissampelos Pareira. Brazil.

Characters.—Cylindrical oval or compressed pieces, entire or split longitudinally, half an inch to four inches in diameter, and four inches to four feet in length. Bark greyish-brown, longitudinally wrinkled, crossed transversely by annular elevations; interior woody, yellowish-grey, porous, with well-marked often incomplete concentric rings and medullary rays. Taste at first sweetish and aromatic, afterwards intensely bitter.

Composition.—Pareira root contains, amongst other ingredients, an active principle, *pelosin*, believed to be identical with beberia.

Incompatibles.—Persalts of iron, salts of lead, and tincture of iodine.

Preparations.

1. **Decoctum Pareiræ.**—1 in 13⅓.   *Dose*, 1 to 2 fl.oz.

2. **Extractum Pareiræ.**—Aqueous. 16 in 1. *Dose*, 10 to 20 gr.

3. **Extractum Pareiræ Liquidum.**—1 in 1.   *Dose*, ½ to 2 fl.dr.

ACTION AND USES.

The physiological action of Pareira is imperfectly known, but it is believed to possess mild bitter and **laxative** effects, and to be a moderately active **diuretic**.

Empirically, it is used in inflammatory affections of the urinary tract, from the pelvis of the kidney downwards, being held to relieve pain, reduce irritation, and promote healing and cessation of muco-purulent discharge. The extract is given along with the decoction to increase its strength; not alone.

Cocculus Indicus.   (*Not Officinal.*)—The fruit of Menispermum cocculus, the Cocculus indicus plant. From the East Indies.

Characters.—A small dark brown berry containing a yellowish reniform seed.

Composition.—The active principle of cocculus is a bitter neutral substance, picrotoxine, $C_9H_{10}O_4$, in colourless crystals, neutral, soluble with difficulty in water. It is united with *menispermic* or *cocculinic acid*, and other principles.

Dose of Picrotoxin.—$\frac{1}{120}$ to $\frac{1}{20}$ gr.

ACTION AND USES.

Externally, cocculus or picrotoxin, in the form of a dilute ointment, very carefully applied to the unbroken surface, destroys **pediculi**.

Internally, picrotoxin is a very powerful agent, especially stimulating the spinal cord and medulla, and causing violent **spasms of the flexors**, and intoxication in large doses. It has been chiefly used in the **night-sweating of phthisis**, and in chronic nervous diseases.

PAPAVERACEÆ.

Papaveris Capsulæ — Poppy Capsules. — The nearly ripe dried capsules of the White Poppy, Papaver somniferum. Cultivated in Britain.

Characters.—Globular, two or three inches in diameter, crowned by a sessile stellate stigma.

Composition.—Poppy capsules contain a little *opium* and woody fibre; the seeds a bland oil. See *Opium.*

Preparations.

1. **Decoctum Papaveris.**—1 in 10.

2. **Extractum Papaveris.**—Aqueous. 3 in 1. *Dose*, 2 to 5 gr.

3. **Syrupus Papaveris.**—1 in nearly 2¼. *Dose*, 1 fl.dr.

ACTION AND USES.

The action of poppy capsules is the same as that of opium, but much weaker. The warm decoction is a favourite **anodyne** fomentation. The extract and syrup are uncertain remedies, and opium preparations are in every respect preferable.

Opium—Opium.—The juice, inspissated by spontaneous evaporation, obtained by incision from the unripe capsules of the Poppy, Papaver somniferum. Grown in Asia Minor,

Characters.—Irregular lumps, weighing from four ounces to two pounds; enveloped in the remains of poppy leaves, and generally covered with the chaffy fruits of a species of rumex; when fresh, plastic, tearing with an irregular slightly moist chestnut-brown surface, shining when rubbed smooth with the finger, having a peculiar odour and bitter taste.

Test.—This is a modification of the process for making hydrochlorate of morphia. (*See* page 186.) 100 gr. of opium ought to yield at least 6 to 8 gr. of morphia.

Varieties.—There are two varieties of officinal opium, Smyrna opium, and that of Constantinople. 1. *Smyrna, Turkey,* or *Levant* opium is the best. It occurs in irregular rounded or flattened masses, seldom more than two pounds in weight, enveloped in poppy leaves, and surrounded with the fruits or seeds of a species of rumex. Good Smyrna opium yields about 8 per cent. of morphia. 2. *Constantinople* opium is of very uncertain quality, generally inferior to Smyrna. **It** is found in cakes, either large and irregular, or small and lenticular, covered with a poppy leaf, and marked with its mid-rib, but without rumex seeds. It smells much less strongly than Smyrna opium. Besides the two officinal varieties, there are found in the market Egyptian opium, in round flattened cakes of a reddish hue, with vestiges of a leaf; Persian opium in sticks or lumps; Indian opium in balls, enveloped in poppy leaves, or in cakes; and French and English varieties.

Impurities (chiefly adulterations).—Opium is often soft from excess of water, which causes great variation in the strength. Stones, fruits, leaves, etc., may be detected by filtering a decoction; and starch by the iodine test. The officinal test is intended to ascertain the amount of morphia in specimens which are pure but of doubtful richness.

Composition.—Opium contains (1) certain *alkaloids*; (2) two *neutral substances*; (3) two organic *acids*; (4) water, resin, gum, extractives, odorous principles, and other constituents of plants. The important components are as follows :

	Parts in 100 Parts.	Constitution.	Reaction.	Characters.
1. Morphia	5 to 20	$C_{17}H_{19}NO_3$	Alkaline	See below.
2. Codeia	up to ·6	$C_{18}H_{21}NO_3$	Alkaline	White octahedra or rhombic prisms.
3. Thebaia or Paramorphia	up to ·3	$C_{19}H_{21}NO_3$	Alkaline	White plates, with acrid styptic taste.
4. Opianin ...		—	—	
5. Cryptopia	·5 to 1	$C_{23}H_{25}NO_5$	Alkaline	
6. Metamorphia		—	—	
7. Papaverina		$C_{20}H_{21}NO_4$	Alkaline	White needles.
8. Narcotin ...	4 to 6	$C_{22}H_{23}NO_7$	Alkaline	Shining prisms; tasteless, odourless.
9. Narcein	upto·02	$C_{23}H_{29}NO_9$	Neutral	Fine white needles; odourless, bitter.
10. Porphyroxin...	—	—	—	
11. Laudanin ...	—	$C_{26}H_{25}NO_3$		
12. Meconin	·08 to ·3	$C_{10}H_{10}O_4$	Neutral	White needles; odourless, acrid.
13. Meconic Acid	4 to 8	$C_7H_4O_7$	Acid	White crystalline pearly scales.
14. Thebolactic Acid	—	Probably Lactic Acid	Acid	—

General chemical characters, Reactions, and Incompatibles.—A fluid (watery or spirituous) preparation of opium reddens litmus

paper (free meconic acid) ; gives a deep red colour with perchloride of iron (meconic acid) ; precipitates with acetate and
subacetate of lead, nitrate of silver, zinc, copper, and arsenic
(meconates, sulphates, and colouring matter) ; a precipitate
with tincture of galls or astringent preparations (tannates of
morphia and codeia) ; and becomes turbid with fixed alkalies,
and the carbonates, alkaline earths, and ammonia (precipitated
morphia and narcotin).

Dose.—$\frac{1}{2}$ to 2 gr.

Preparations.

1. **Emplastrum Opii.**—1 in 10.
2. **Extractum Opii.**—Aqueous. <u>2 of opium in 1.</u> *Dose*, $\frac{1}{2}$ to
    1 gr.

    From Extractum Opii are prepared :
        a. **Extractum Opii Liquidum.**—Made by digesting
    the Extract in water, and adding spirit. 1 of opium,
    i.e. $\frac{1}{2}$ of extract, in 10. *Dose*, 10 to 40 min.
        b. **Trochisci Opii.**—$\frac{1}{10}$ gr. of Extract in each. *Dose*,
    1 to 2.
        c. **Vinum Opii.**—$\frac{1}{2}$ of Extract, *i.e.* 1 of opium, in
    10 of Sherry, with Cinnamon and Cloves. *Dose*, 10 to
    40 min.

3. **Pilula Plumbi cum Opio.**—Opium, 1 ; Acetate of Lead, 6 ;
    Confection of Roses, 1. <u>1 in 8.</u> *Dose*, 4 to 8 gr.
4. **Pilula Saponis Composita.**—Opium, 1 ; Hard Soap, 4 ;
    water, q.s. <u>1 in 6.</u> *Dose*, 3 to 5 gr.
5. **Pulvis Opii Compositus.**—Opium, 3 ; Black Pepper, 4 ;
    Ginger, 10 ; Caraway, 12 ; Tragacanth, 1. <u>1 in 10.</u>
    Dose, 2 to 5 gr.

    From Pulvis Opii Compositus is prepared :
        a. **Confectio Opii.**—Compound Powder in Syrup.
    <u>1 of opium in 40.</u> *Dose*, 5 to 20 gr.

6. **Pulvis Ipecacuanhæ Compositus.**—Dover's Powder. Opium,
    1 ; Ipecacuanha, 1 ; Sulphate of Potash, 8. <u>1 in 10.</u>
    Dose, 5 to 15 gr.

    From Dover's Powder is prepared :
        a. **Pilula Ipecacuanhæ cum Scillâ.** — Compound
    Ipecacuanha Powder, Squill, Ammoniacum, and Treacle.
    <u>1 of opium in 23.</u> *Dose*, 5 to 10 gr.

7. **Pulvis Kino Compositus.**—Opium, 1 ; Kino, 15 ; Cinnamon,
    4. <u>1 in 20.</u> *Dose*, 5 to 20 gr.
8. **Pulvis Cretæ Aromaticus cum Opio.**—Opium, 1 ; Aromatic
    Chalk Powder, 39. <u>1 in 40.</u> *Dose*, 10 to 40.

9. **Suppositoria Plumbi Composita.**—Opium, Acetate of Lead, Benzoated Lard, White Wax, and Oil of Theobroma. 1 gr. in each.

10. **Tinctura Opii.**—"Laudanum." Opium, 1½; Proof Spirit, 20. 1 gr. in 14⅔ min. *Dose*, 5 to 40 min.

From Tinctura Opii are prepared .

 a. **Enema Opii.**—Tincture, ½ fl.dr.; Mucilage of Starch, 2 oz. For one enema.

 b. **Linimentum Opii.**—Equal parts of Tincture and Soap Liniment.

11. **Tinctura Opii Ammoniata.**—"Scotch Paregoric." Opium, Saffron, Benzoic Acid, Oil of Anise, Strong Solution of Ammonia, and Spirit. 1 in 96. *Dose*, ½ to 1 fl.dr.

12. **Tinctura Camphoræ Composita.**—Opium, Benzoic Acid, Camphor, Oil of Anise, Proof Spirit. ¼ gr. in 1 fl.dr. *Dose*, 15 to 60 min.

13. **Unguentum Gallæ cum Opio.**—Opium and Ointment of Galls. 1 in 14⅔.

From Opium is made :

Morphiæ Hydrochloras.—Hydrochlorate of morphia. $C_{17}H_{19}NO_3.HCl.3H_2O$.

 Source.—Made by (1) precipitating and rejecting the meconic acid and resins, by adding a solution of chloride of calcium to a concentrated cold watery infusion of opium; (2) evaporating the solution (containing hydrochlorates of the alkaloids), pressing to remove colouring matter, exhausting with boiling water, filtering, and pressing again; (3) repeating process (2) until solution is nearly colourless; (4) completing decolorisation by digesting with charcoal and filtering; (5) precipitating morphia by ammonia, washing, diffusing in water, dissolving in hydrochloric acid and crystallising out.

 Characters.—White acicular prisms of silky lustre; soluble in water and in spirit.

 Incompatibles.—The alkaline carbonates, lime-water, salts of lead, iron, copper, mercury, and zinc, liquor arsenicalis, and all astringent vegetables.

 Dose.—1/16 to ¼ gr.

a. **Liquor Morphiæ Hydrochloratis.**—Solution of Hydrochlorate of Morphia. 4 gr. in 1 fl.oz. of a mixture of Spirit, Water, and Diluted Hydrochloric Acid. *Dose*, 10 to 60 min.

b. **Suppositoria Morphiæ.**—½ gr. in each.

c. **Suppositoria Morphiæ cum Sapone.**—½ gr. in each.

d. **Trochisci Morphiæ.**— $\frac{1}{36}$ gr. in each.

e. **Trochisci Morphiæ et Ipecacuanhæ.**— $\frac{1}{36}$ gr. with $\frac{1}{12}$ gr.
Ipecacuanha in each.

From Morphiæ Hydrochloras is made:

Morphiæ Acetas.—Acetate of Morphia. $C_{17}H_{19}NO_3$. $C_2.H_4O_2$.

Source.—Made by precipitating morphia from a solution of the hydrochlorate by means of ammonia, dissolving in acetic acid and water, and evaporating.

Characters.—A white powder, soluble in water and in spirit.

Dose.—$\frac{1}{8}$ to $\frac{1}{2}$ gr.

α. **Injectio Morphiæ Hypodermica.**—Hypodermic Injection of Morphia. 1 gr. acetate in 12 min., made by freshly preparing the acetate as above, but without evaporating. *Dose,* hypodermically, 1 to 6 min.

β. **Liquor Morphiæ Acetatis.**—Solution of Acetate of Morphia. 4 gr. in 1 fl.oz. of Spirit, Water, and Diluted Acetic Acid. *Dose,* 10 to 60 min.

ACTION AND USES.

1. IMMEDIATE LOCAL ACTION AND USES.

Externally.—Opium is very generally believed to be anæsthetic and anodyne when applied to the unbroken skin, and the emplastrum, linimentum, fomentations, and other preparations are used to relieve the pains of neuralgia, lumbago, abscess, etc. It is doubtful, however, whether morphia can be absorbed by the unbroken skin, and the benefit derived from these applications may be referable to the spirit, resins, and heat. Wounds, ulcers, and exposed mucous surfaces readily absorb opium, which is used in painful ulcers, conjunctivitis, and similar diseases. It is occasionally given by the endermic method, especially in the epigastric region. Hypodermic injection is a most valuable means of administering morphia, when a specially rapid or local effect is desired, or when the stomach is irritable or inaccessible.

Internally.—Opium has a peculiar taste, is quickly absorbed by the mucous membrane, and exerts an action upon the mouth, which, although in part specific and in part remote, is chiefly an immediate local one. A full medicinal dose renders the mouth dry and the tongue foul, from **diminution of the secretions**, with thickness of the voice and some thirst. On entering the stomach opium may cause sickness, from brief irritation of the nerves, but sensibility is quickly reduced, hunger and pain relieved or removed; appetite, gastric secretion, and digestive activity diminished; and the afferent impressions which give rise to vomiting arrested, so that direct emetics will no longer act. Anorexia, nausea, and sickness may occur as *sequelæ* of the same or larger doses.

These effects of opium on the stomach have a double bearing in therapeutics. First, they indicate that it has a constant tendency to **derange digestion.** Secondly, it is a powerful means of relieving gastric pain and vomiting, whatever their cause, but especially in the acute catarrh which remains as the effect of irritant food, alcohol, or poison, after these have been removed; in ulcer, "chronic," or malignant; and in reflex sickness, due to disease, irritation, or operation, in some other part of the abdomen. In chronic dyspeptic pain it is manifestly contra-indicated.

The action of opium on the intestine is distinctly sedative, although very brief primary stimulation may sometimes be recognised. Both the sensible and insensible impressions from the mucous membrane are diminished or arrested by medicinal doses. Pain is prevented or relieved, the secretions are less abundant, and peristalsis is more feeble or arrested; the total result being **anodyne** and **astringent.** Opium is therefore a most valuable remedy for unnatural frequency of the bowels, as in simple diarrhœa, dysentery, the first stage of cholera, the ulceration of typhoid fever and tuberculosis, and irritant poisoning. In all such cases, however, it must be employed with the cautions to be afterwards insisted on, and in most instances it is best prescribed as an addition to other astringents such as chalk, lead, and tannic acid in its many forms; the amount of opium being a minimum, but still sufficient to assist the less powerful drugs. It has the further advantage of relieving abdominal pain. Even infants (see *cautions*, page 197) may thus be treated for diarrhœa with the greatest benefit.

Opium is of still greater service in **paralysing the bowels** in hernia, intestinal obstruction, peritonitis, and visceral perforations, ruptures, and wounds. The drug must be freely and continuously given in such cases, until nature or art can afford relief.

Given by the rectum, as the enema or suppository, opium relieves local pain, diarrhœa, dysentery, and spasm of the rectum or neighbouring parts, sets the pelvic organs at rest after operations, and prevents irritability of the rectum by nutrient enemata. The dose of opium by the rectum should be half as much more as by the mouth. A trace of morphia is excreted unabsorbed in the fæces.

2. ACTION ON THE BLOOD.

Morphia enters the circulation less quickly than some other alkaloids, although the first traces of the drug are rapidly discovered in the blood. Thus its full action is comparatively slowly developed, and solid opium continues to exert local effects even in the colon, portion by portion of the morphia being absorbed into the vessels. The red corpuscles are said to be reduced in size indirectly, possibly through slowing of the circulation and want of oxygen.

3. SPECIFIC ACTION.

After administration morphia may be found in all the organs, which, probably without exception, are physiologically affected by it; but its principal action is exerted upon the nervous system.

The *convolutions* are first briefly excited, and afterwards depressed, probably by direct action of the morphia upon the nerve-cells, not on the cerebral vessels. The stage of opium excitement transcends even the first stage of alcoholic intoxication in the exaltation of feelings, the sense of happiness and comfort, the brilliancy of imagination, and the increase of intellectual power and mental vigour generally, all accompanied by brightness of expression and manner. But the effect of opium, even in this stage, is rarely one of pure exaltation, and in most persons is perhaps never so. There is generally some perversion of the faculties, and the imagination becomes extravagant, wandering into the land of dreams, of the grotesque, and the impossible. Depression now supervenes: the various perceptive and sensory centres in the convolutions are more or less depressed, according to the dose; impressions made upon the afferent nerves, including pain, do not readily affect the receptive centres; the subject becomes drowsy, and finally sleeps; and if he momentarily respond to a sharp enquiry or other forms of stimulation, he quickly relapses into heavy sopor. If the dose has been excessive, the stage of excitement is entirely absent, the cerebrum is speedily and profoundly depressed, and no response follows severe forms of stimulation, such as flagellation: the patient is comatose. These effects of opium on the brain as a **stimulant, hypnotic, anodyne, and narcotic,** are more marked in man and in highly intellectual races than in animals and lower races respectively. In cold-blooded animals they are quite subordinate to the effects of stimulation of the cord.

The *ganglia at the base of the brain* are affected by opium, causing **contraction of the pupil,** and disturbance of accommodation.

The grey matter of the *spinal cord* is at first, and briefly, stimulated by a moderate dose of opium, reflex excitability being increased, as shown by restlessness in man and convulsions in animals. At the stage of cerebral depression, languor and muscular weakness, of spinal origin, set in, and the subject lies down; but there is not even then complete loss of muscular power and irritability; and even in dangerous poisoning by opium the patient can be marched about, if supported on either side.

Following close upon the convolutions and cord, the great vital centres in the *medulla* are markedly affected by opium. *Vomiting* is not uncommon as one of the first effects. The *respiratory* centre, at first unaffected, is then depressed, the respiratory movements becoming quiet, superficial, and irregular; and death by opium poisoning is due to **paralysis of the respiratory centre** and arrest of breathing, that is, to asphyxia. The *cardiac* centre is more resistant to morphia, and is first excited so as to increase inhibition (after an evanescent acceleration); but it is soon depressed, and the pulse rises in frequency. The *vascular* centre is depressed by opium, but never to a dangerous extent; and even in complete narcosis, when respiration is failing, the blood pressure (pulse) responds to afferent stimuli. The full action of opium on the respiration, heart, and vessels will be immediately described.

We shall presently find that the therapeutical value of the action of opium on the central nervous system lies in the fact that it depresses the perceptive and sensory centres so much earlier and more profoundly than the vital centres in the medulla. Its effect on the pupil, heart, vessels, respiration, and cord are either of little positive value in treatment, or are altogether unfortunate.

The functions of the *sensory nerve terminations* are lowered or arrested by opium, common sensibility being especially reduced, so that pain cannot be originated; but this peripheral anæsthetic or anodyne effect of morphia given by the mouth is decidedly secondary, both in time and in degree, to its allied action on the sentient centres, and to its local effect when administered by hypodermic or interstitial injection, already described.

The *sensory nerve-trunks* are diminished in conductivity by local injection of morphia, as well as by its internal administration, thus offering a second interruption to the flow of painful impressions inwards.

The *motor nerves* are first briefly excited, and then paralysed from the centre outwards. *Muscular* irritability is never completely lost.

The action of opium upon the centres of several of the *viscera* has been partly described under the previous heads. In addition to this, it **depresses the afferent** (including the sensory) **nerves of all organs**, and acts upon many of the viscera directly.

The *heart* is temporarily accelerated by opium, in part through the cardiac centre, in part through its intrinsic ganglia. Thereafter, or with fuller doses, it is slowed by stimulation of the vagus in the medulla and heart. Finally, the **cardiac vagus is depressed** or paralysed; but by this time the intrinsic ganglia are so depressed that acceleration is impossible, and the action remains infrequent, whilst very feeble. Very rarely death occurs by sudden cardiac failure.

The vessels, dilated through the centre, as described, are not directly influenced by opium, either in their muscular coats, or in their peripheral nerves.

Whilst the *respiratory* movements of the chest are impaired through the centre, so that they become feeble and tend to cease, the afferent nerves of breathing—that is, the branches of the vagus arising in the lungs and passages—are also depressed. Thus reflexion is dulled or arrested at its very origin, and dyspnœal excitement (hyperpnœa), cough, spasm, and other reflex respiratory acts are rendered more difficult or altogether prevented. At the same time, the bronchial secretions are diminished or inspissated by the action of the drug upon the glands, and the activity of pulmonary circulation is lowered with the general blood pressure, and by the weakening of the respiratory movements. The total effect of opium upon the respiratory functions is thus powerfully depressant.

The biliary and glycogenic functions of the *liver* are affected by morphia, which causes pale stools or even jaundice, and remarkably diminishes the amount of sugar in diabetes. **Hepatic and general metabolism is reduced in activity**, the amount of urea and probably of carbonic acid excreted being distinctly diminished. The *temperature* rises for a time, and then falls, apparently varying with the blood pressure.

4. SPECIFIC USES.

The hypnotic and anodyne effects of opium constitute it by far the most valuable drug of its kind, and the most important article of the whole *materia medica*. It is constantly employed to induce sleep, relieve pain, and calm excitement; this combination of properties giving opium a great superiority to chloral and other simple hypnotics, on the one hand, and to aconite, belladonna, quinine, and other direct or indirect anodynes, on the other hand. Speaking broadly, it is used in sleeplessness due to pain; in the insomnia of exhaustion, over-work, fever, or insanity; and in the restlessness and anxiety of visceral disease; the quantity, combinations, and time of administration being carefully arranged. In delirium chloral is often preferred, especially in delirium tremens; but opium is more suitable in the delirium of mania, and in the later stages of fevers, when the temperature is falling and the respiration and circulation are not oppressed. It has been recommended, however, in heat-pyrexia, combined with quinia.

There are but few kinds of pain that cannot be relieved by opium; but whether it be wise to administer it in every instance is another question. The unbearable pains attending the passage of renal and biliary calculi, the pains of neuralgia, acute rheumatism, and cancer; of fractures, dislocation, and other injuries, are a few examples of conditions in which opium is essential. In all cases when pain is urgent, and its seat accessible, the hypodermic method should be chosen. In gout it is to be used only when the pain is excessive, as it tends to aggravate the cause. In hysterical pain it is less valuable. Other local visceral pains will be noticed presently. The pain and shock of operations are constantly treated with a full dose of opium.

No use is made of the action of opium on the iris and ciliary nerves.

As an antispasmodic, opium is less employed for various reasons, *e.g.* in epilepsy and other convulsive diseases; but it relieves some cases of spasmodic asthma, whooping-cough, and spasmodic stricture of the urethra.

The violent spasms and pains of certain diseases of the *cord* may yield to no other form of treatment than morphia injected hypodermically.

From its action on the medulla, opium has been recommended as an antidote to belladonna, which is so far its physiological antagonist, as we shall see (page 198); but it must always be used with great caution, and only in the stage of excitement.

The practical points connected with the vital centres will be noticed under the heart, vessels, and respiration.

In disease of the *heart*, opium is of great value to relieve pain, anxiety, and distress, whilst, as we have seen, it is a dangerous cardiac depressant. Towards the end of most cases of cardiac disease, the greatest discrimination is called for as to when opium may or may not be given. The safe rule is to trust to other anodynes entirely, such as belladonna, direct and indirect stimulants, and measures for relieving the circulation; but it is equally true that in some cases of heart disease unspeakable relief and permanent benefit may be obtained by the hypodermic injection of morphia. This subject must be studied in books on the practice of medicine.

From its soothing effect upon the *vessels* and circulation generally, opium is a hæmostatic of the first order, but requires to be used with judgment. In hæmoptysis, it is given in small doses, to relieve cough, to depress the circulation slightly by slowing and weakening the heart and dilating the vessels, and to relieve the mind of the anxiety which aggravates the bleeding. In intestinal hæmorrhage it is of great value, arresting, as it also does, the movements of the bowel. It is best given combined with lead or preparations containing tannic acid.

The soothing influence of opium on the bronchi, lungs, the afferent nerves, and the centre of *respiration*, accounts for its extensive employment in cough, pain, dyspnœa, and other distressing symptoms in the chest. Its power here is unquestionable ; but for this very reason the danger of it is great. Cough and dyspnœa are frequently beneficial acts, and are not to be arrested in a routine fashion by sedatives, but, if possible, by the removal of their cause. When cough is due to some irremovable condition, such as growth in the lung or bronchi, pressure, or a remote (reflex) irritation, or to excessive irritability of the nerves and centre, opium is indicated, and may be given with benefit. On the other hand, in cough and respiratory distress with abundant secretion, as in the bronchitis of the old and infirm or of the very young and feeble, opium leads to retention and inspissation of the products, aggravation of the cause, and asphyxia, and is on no account to be given. Between these extremes lies every variety of case in which opium may suggest itself, *e.g.* in phthisis and recurrent bronchial catarrh. The rule here should be on no account to prescribe opium unless other means have failed, such as the many expectorants, and attention to food, warmth, etc.; and that, when given, opium must be ordered in small doses combined with expectorants, such as ammonia and ipecacuanha, which will prevent dangerous depression of the local nerves and centres. In acute inflammation of the pleura, or pleuropneumonia, it may be necessary to relieve severe pain in the chest, harassing cough, sleeplessness, and mental distress by morphia hypodermically. For asthma, opium must be ordered with the greatest hesitation, as the opium habit is readily acquired in this disease. Its employment in hæmoptysis has been already noticed.

With respect to the *liver and metabolism*, opium is by far the most powerful drug known in reducing or removing sugar from the urine in diabetes, and therewith ameliorating the condition of the patient in most respects. Very large doses of

N—8

solid opium, morphia, or better still, codeia, may be tolerated in this disease, their effect on the nervous system being remarkably absent whilst the diabetes is yielding. Acute inflammatory and febrile diseases are now less frequently treated with opium than formerly, when a combination with calomel was generally used, the opium preventing the purgative action of the mercurial, and the latter preventing constipation, whilst both drugs were believed to act specifically on the morbid process, reducing the local general circulation, alleviating pain and restlessness, and promoting healing. The combination is, however, very valuable in syphilis. In the specific fevers, such as typhoid, opium given with judgment relieves delirium, as we have seen, checks diarrhœa, and is invaluable in hæmorrhage, perforation, or peritonitis. With quinia it is given in some cases of malaria. Phagedæna may call for its free exhibition.

Opium is employed in obstetrics to prevent abortion, in some varieties of difficult labour, and to relieve after-pains.

5. REMOTE LOCAL ACTION AND USES.

The excretion of morphia commences quickly, but may not be completed for forty-eight hours. It passes out of the body by most of the secretions, especially the urine, where it is found mainly unchanged. The **quantity of urine is diminished**; its evacuation sometimes disturbed or difficult, from the local action of morphia on the bladder; and sugar occasionally present. These facts, and the probability of the retention and accumulation of morphia in the system if the action of the kidneys be deficient, indicate the necessity to give it only with the greatest caution, in reduced doses, or not at all, in renal disorder or disease.

Opium in passing through the *skin* may cause itching, heat, and sometimes eruptions. The vessels are also dilated, as we have seen, and the sweat glands decidedly stimulated; both being effects of its central, not local, cutaneous action. Thus opium, especially in the form of Dover's powder, is a valuable **diaphoretic,** and is given with great success as a refrigerant apyretic in the outset of catarrh, influenza, and mild febrile or rheumatic attacks caused by cold. Under certain circumstances, Dover's powder actually checks the sweating of phthisis, probably by removing its cause. Being excreted in the milk, opium must be prescribed with caution to nursing females.

6. ACTION AND USES OF THE PRINCIPAL CONSTITUENTS OF OPIUM.

1. **Morphia.**—The **action of opium depends** chiefly **on morphia,** and the description just given applies so nearly to the

pure alkaloid, that only a few points of difference require to be noticed. These depend upon two principal circumstances: (1) Opium, being much less soluble than the pharmacopœial preparations of morphia, is more slowly absorbed, and thus acts less quickly than morphia, whilst its effects are more lasting, and its immediate local action on the intestines decidedly more marked. (2) Several of the constituents of opium possess more or less convulsant action (thebain, codeia, narcotin), morphia none (in man); the latter has therefore a somewhat more sedative influence than the entire drug. The effect of opium on the skin is also less marked in morphia. Unless there be some special reason to the contrary, morphia is generally to be preferred to opium in practice, as being of definite composition (whilst the crude drug is very variable), more rapid in action, readily administered hypodermically, whilst the dyspeptic and constipating effects of the drug are less marked. Opium is to be preferred in intestinal and abdominal diseases such as diarrhœa, obstruction, peritonitis, hernia, because it reaches the bowel; in delirium tremens and mental disorder, because its action is more continued; in diabetes, because it contains codeia; for combinations with quinine or calomel, and as a diaphoretic, because it prevents purgation and lowers fever; in astringent enemata, from its action on the bowel; and for local applications, *e.g.* to the conjunctiva, because less irritant than the alkaloid. The relative strength of opium to morphia is about 20 or 30 to 100, $\frac{1}{5}$ or $\frac{1}{3}$ to 1.

2. *Codeia.*—This alkaloid appears to **excite the cord** more than morphia, and to depress the convolutions less, so that muscular tremors may follow and exceed the sedative influence. Codeia, in $\frac{1}{2}$-gr. doses cautiously increased, markedly **reduces the amount of sugar** in diabetes, appearing to act as an alterative to the nervous system, and thus to cure (not simply relieve) the disease in some instances.

3. *Narcotin*, which is so large a constituent of opium, is probably often impure from an admixture of morphia. By some authorities it is considered to be **hypnotic,** by others convulsant. It is not used.

4. *Narceïn* probably **acts like morphia,** and is not employed medicinally.

5. *Thebain* is a **convulsant,** almost like strychnia, but is not used.

6. Opianin, cryptopia, metamorphia, and possibly papaverina, **act like morphia.** Porphyroxin and laudanin **act like codeia.**

The action of meconic acid is doubtful.

7. APPLICATIONS OF THE VARIOUS PREPARATIONS OF OPIUM.

This subject will be best discussed from the point of view of the conditions calling for opium.

1. *Severe pain*, such as colic or neuralgia, is to be treated with the Hypodermic Injection of Morphia. Failing this, either of the Solutions of Morphia must be given by the mouth, or a fluid preparation of opium, such as the Tincture, or Liquid Extract (about ⅓ more active than the tincture). The Enema is a valuable anodyne in cases of abdominal pain. The Pilula Saponis Composita also acts rapidly, being more readily soluble in the stomach than solid opium.

2. *Superficial pain* may be met by local applications such as the Plaster, Liniment, or fomentations made with laudanum or other fluid preparation; but, as we saw, the value of the drug itself in all these applications is very doubtful.

3. *As a hypnotic*, the best forms are the Tincture, the Liquid Extract, the Solution of Morphia, and the Soap and Opium Pill; the particular preparation and the dose being regulated by the degree of sleeplessness and of the pain which may accompany it. Dover's Powder is an excellent hypnotic in the restlessness at the commencement of feverish attacks.

4. As a *sedative to the stomach*, various preparations may be tried, such as the Solutions of Morphia in effervescing mixtures, morphia endermically or hypodermically over the epigastrium; sometimes solid opium or the Extract in the form of a small pill. Dover's Powder is of great value in painful ulceration and acute dyspepsia, combined with bismuth or soda.

5. As a *sedative and astringent to the bowels*, Laudanum, either by the mouth or as the Enema, may be given in urgent cases attended by much pain. When there is less urgency, we may prescribe one of the powders—Compound Opium Powder, Chalk and Opium, Kino and Opium, or Dover's Powder. Acetate of morphia with acetate of lead and acetic acid, or the Lead and Opium Pill may be demanded in severe diarrhœa, especially if hæmorrhage threaten. Solid opium, alone or combined with calomel, is the best form of astringent when the bowel must be paralysed, as in hernia, peritonitis, and intestinal obstruction.

6 As *a sedative to the rectum*, bladder, pelvic organs, and urethra, we possess the various Suppositories of opium and morphia, and the Enema.

7. *Cough* may be relieved by several special preparations, namely: Tinctura Camphora Composita, Tinctura Opii Ammoniata, the three Trochisci, and the Pilula Ipecacuanhæ cum Scilla.

8. *Diaphoresis* is generally accomplished with Dover's Powder.

The uses of the other preparations are obvious. The Confection is a pleasant form of the compound powder.

Influences modifying the action and uses of opium.— Dangers : Cautions.—*Age* modifies greatly the effects of opium, children being particularly susceptible of its influence on the convolutions and medulla. An infant of one year should not be given more than half a minim of the tincture for an ordinary dose, and suckling women should be ordered opium with special precautions. *Females* are more easily affected than males. Certain individuals have peculiar *idiosyncrasies* as regards opium, some resisting its action, others being excited by it, others again very readily narcotised; whilst more frequently some persons suffer from a species of shock after the hypodermic injection of morphia, becoming sick, faint, and even alarmingly collapsed. The effect of *habit* is extremely marked in opium, the necessary dose steadily rising, until large quantities may be safely taken. *Disease*, especially *pain*, affords great resistant power to the action of opium, which appears to expend its action on the morbid process. The quality of the opium, the particular preparation, and the combinations used, also modify its action. On the contrary, opium and morphia act more powerfully in the subjects of kidney disease, as we have already seen. Morphia and opium are *contra-indicated*, because dangerous, or are to be used with special care, in diseases of the respiratory organs, the heart, and the kidneys; in congestive conditions and hyperæmia of the brain; and in alcoholic intoxication.

Opium and Belladonna : Combinations and Antagonism of Morphia and Atropia.—In several respects the action of morphia is opposed to that of atropia, the important principle of belladonna. The *antagonism* between the two substances is in part real, such as their respective effects on the convolutions, respiratory centre, and intestines; in part apparent only. Thus, the contraction of the pupil caused by morphia occurs through the basal ganglia; the dilatation caused by atropia is referable to paralysis of the ciliary branches of the third nerve. Morphia is a diaphoretic through the centres : atropia an anhidrotic through the terminal nerves of the glands. Both depress the heart and reduce the blood pressure, in poisonous doses. Thus, morphia and atropia are not true antagonists, but the one may prevent or relieve certain effects of the other, and may therefore be combined with the other for particular medicinal purposes, or given in the treatment of poisoning by the other under particular circumstances. *Combinations* of atropia and morphia are now extensively used for hypodermic injection ($\frac{1}{100}$, $\frac{1}{50}$, or even $\frac{1}{25}$ gr. of sulphate of atropia to 1 gr. of morphia), to prevent certain unpleasant effects of the latter. It is found that the immediate sickness and depression, and the subsequent dyspepsia and constipation, may thus be avoided, and a more natural sleep induced. The combination is preferable when morphia is given as a hypnotic or anodyne; in conditions of cardiac depression and disease of the lungs; in obstruction of the bowels; and to relieve spasms in general. The atropia should be avoided in cerebral excitement, especially mania.

Use as mutual antidotes.—Sulphate of atropia, in doses of $\frac{1}{100}$ gr., may be injected subcutaneously every quarter of an hour in opium poisoning, the pulse and respiration being carefully watched. Three or four doses may thus be given; but the ordinary means of resuscitation, especially artificial respiration, must not be for a moment interrupted.

In poisoning by belladonna, morphia should be given subcutaneously, with the same precautions, in doses of $\frac{1}{4}$ of a grain.

Apomorphia. $C_{17}H_{17}NO_2$. APOMORPHIA. (*Not Officinal.*)

Source.—Made by heating morphia in a closed tube, with concentrated hydrochloric acid, whereupon the alkaloid loses one molecule of water—$C_{17}H_{19}NO_3 = C_{17}H_{17}NO_2 + H_2O$.

Characters.—A white powder, becoming green on exposure or in solution, without loss of its properties. Soluble in ether and alcohol. The hydrochlorate of apomorphia, which is generally used, occurs as minute greyish crystals, soluble in water. Solutions should be freshly prepared for use.

Dose of the hydrochlorates.—$\frac{1}{10}$ to $\frac{1}{8}$ gr., by the mouth; $\frac{1}{20}$ to $\frac{1}{10}$ gr., hypodermically.

ACTION AND USES.

Apomorphia is the most certain of all emetics, acting upon the vomiting centre, and not on the stomach, *i.e.* being an **indirect emetic.** In from five to twenty minutes it induces moderate nausea, repeated vomiting, and the disturbances of the respiratory and circulatory organs, characteristic of this class of remedies. If the dose have been sufficient, the evacuation of the stomach is certain and complete. Larger doses cause prostration and paralysis of the voluntary muscles, depression of the respiratory centre, acceleration of the heart, and fall of temperature. Small doses are **expectorant.** Apomorphia may be used for the

many purposes of emetics in general. Its special advantages
consist in its certainty; the absence of local irritation of the
stomach; the readiness with which it can be given hypodermically, that is, to patients unable to swallow, as a small non-
irritant injection; and the absence of after-effects. Its
expectorant action has been but little employed.

Rhœados Petala—Red-Poppy Petals.—The
fresh petals of Papaver Rhœas. From indigenous
plants.

Characters.—Of a scarlet colour and heavy poppy odour.
Composition.—Red poppies contain a large quantity of
colouring matter, readily soluble in water, consisting of two
acids, *papaveric* and *rhœadic acids ;* also an alkaloid *rhœadin,*
$C_{21}H_{21}NO_6$, without narcotic properties. Red poppies contain
no morphia.

Preparation.

Syrupus Rhœados.—1 in $3\frac{1}{2}$. *Dose,* 1 fl.dr.

ACTION AND USES.

Syrup of red poppies is used as a **colouring agent** only.

CRUCIFERÆ.

Sinapis—Mustard.—The seeds of Sinapis Nigra,
and Sinapis Alba; also the seeds reduced to powder,
mixed.

Characters of the Powder.—Greenish-yellow, of an acrid
bitterish oily pungent taste, scentless when dry, but exhaling
when moist a pungent penetrating peculiar odour, very irritating
to the nostrils and eyes. *Impurity.*—Starch ; a decoction cooled
should not be made blue by tincture of iodine.

Substances resembling black mustard : Colchicum seeds,
which are larger, lighter, and not quite round.

Composition.—The seeds of *sinapis nigra* contain : (1) about
35 per cent. of a bland *fixed oil.* When this has been removed
by expression, and the powdered mustard mixed with water
and distilled, there is obtained (2) the officinal *volatile oil, Oleum
Sinapis,* C_4H_5NS. This is a colourless or pale yellow body,
nearly insoluble in water, of intensely penetrating odour,
burning taste, and blistering action on the skin. As the seeds
and powder of the mustard are devoid of these irritant properties, the oil cannot exist ready formed in them, but is

developed by a decomposition of their constituents. On the addition of water to the black mustard, its most important principle, *potassium myronate* or *sinigrin* ($C_{10}H_{18}NKS_2O_{10}$), a compound of potassium with an acid glucoside, *myronic acid*, is broken up by another constituent, *myrosin,* a ferment, into volatile oil of mustard, potassium sulphate, and sugar, thus: $K,C_{10}H_{18}NS_2O_{10} = C_4H_5NS + KHSO_4 + C_6H_{12}O_6$. *Sinapis alba* also contains the fixed oil. It does not, however, yield the volatile oil, but a substance with allied properties, called *sulpho-cyanate of acrinyl*, C_8H_7NSO, by a similar decomposition of its constituents, *sinalbin*, $C_{30}H_{44}N_2S_2O_{16}$ (in place of potassium myronate) and myrosin, thus : $C_{30}H_{44}N_2S_2O_{16} = C_8H_7NSO + C_{16}H_{23}NO_5,H_2SO_4$ (disulphate of sinapin) $+ C_6H_{12}O_6$ (glucose).

Preparations.

1. **Oleum Sinapis.**—The oil distilled with water from the seeds of Sinapis nigra after the expression of the fixed oil. Solubility, 1 in 50 of water; readily in spirit and ether.

From Oleum Sinapis is prepared :

a. **Linimentum Sinapis Compositum.**—1 in 41, with Ethereal Extract of Mezereon, Camphor, Castor Oil, and Spirit.

2. **Cataplasma Sinapis.**—Mustard in powder, linseed meal, and boiling water.
3. **Charta Sinapis.**—Made with guttapercha solution.

ACTION AND USES.

1. IMMEDIATE LOCAL ACTION AND USES.

Externally.—When applied to a limited area of skin mustard acts quickly (1) as a rubefacient and nervous stimulant, causing redness, heat, and severe burning pain. (2) This effect is speedily followed by loss of sensibility in the part to other impressions, and relief of previous pain. (3) The prolonged application of the charta or cataplasm causes vesication by the production of local inflammation. Neighbouring and deeper parts, and viscera in vascular communication or intimate nervous relation with the blistered area, may thus have their circulation relieved. The heart, blood pressure, respiration, and nervous centres generally are stimulated by the first application of mustard to the skin; soothed during the stage of anæsthesia, and relief of pain; and depressed in the third stage, especially if the vesication be severe through neglect. Applied to the whole or a large part of the surface of the skin in the form of a bath, mustard dilates the cutaneous vessels, and thus relieves the blood pressure in the viscera.

In the form of poultice or paper, mustard is extensively used as a readily available, convenient, and rapid means of relieving local pain, stimulating the internal organs, and producing counter-irritation, with evanescent and mild after-effects. It is applied to relieve the pains of muscular rheumatism (lumbago, etc.) ; neuralgia in any part of the body ; the indefinite pains in the chest in chronic disease of the lungs or heart ; and colic, gastralgia, and other forms of distress in the abdomen. As a cardio-vascular and respiratory stimulant, a large sinapism may be applied to the calves or soles in syncope, coma, or asphyxia, whether from disease or from poisoning. The counter-irritant effect of mustard is chiefly used in inflammation of the throat, larynx, bronchi, lungs, pleura, and pericardium ; sometimes in abdominal diseases ; frequently, and with success, in morbid conditions of the stomach, and persistent vomiting from any cause. Diffused through a warm bath it is a popular " derivative " in cerebral congestions, headache, and at the onset of colds and febrile diseases in children. A mustard sitz bath may stimulate menstruation if taken at the period.

Internally.—Mustard produces a familiar pungent impression on the tongue and olfactory organs, a sense of warmth in the stomach, and an increase of relish and appetite. The circulation in the gastric wall is also stimulated, but it is remarkable that the effect of mustard on the circulation in the stomach is much less powerful than that on the skin. In full doses it is emetic, with a rapid stimulant action, and little subsequent depression.

Mustard is used internally chiefly as a condiment. As an emetic, from one to four teaspoonfuls may be given stirred up with a tumblerful of warm water in cases where other emetics are not available, or have failed, especially in poisoning by narcotics such as opium.

2. ACTION ON THE BLOOD, SPECIFIC, AND REMOTE LOCAL ACTION.

The odour of oil of mustard can be detected in the blood. Its specific action is obscure, and never taken advantage of medicinally. Part, at least, of oil of mustard is excreted by the lungs.

Armoraciæ Radix—HORSERADISH ROOT.—The fresh root of Cochlearia Armoracia. Cultivated in Britain.

Characters.—A long cylindrical, fleshy root, half an inch to one inch in diameter, expanding at the crown into several very short stems. It is internally white, and has a pungent taste and smell.

Substances resembling Horseradish: Aconite root, which is short, conical, darker, and causes tingling when chewed.

Composition.—Horseradish yields, along with other constituents, a *volatile oil*, C_4H_5SN, closely allied to the volatile oil of black mustard, and formed, like it, by decomposition of a more complex principle by means of a ferment.

Preparation.

Spiritus Armoraciæ Compositus.—1 in 8, with orange-peel, nutmeg, and spirit. *Dose*, 1 to 2 fl.dr.

ACTION AND USES.

Horseradish has been used in domestic medicine as a counter-irritant, but is most familiar as a pleasant condiment, possessing much the same properties as mustard. The compound spirit is a **flavouring and carminative agent.**

POLYGALACEÆ.

Senegæ Radix—SENEGA ROOT.—The dried root of Polygala Senega. From North America.

Characters.—A knobby root-stock, with a branched tap-root, of about the thickness of a quill, twisted and keeled; bark yellowish-brown; sweetish, afterwards pungent, causing salivation; interior woody, tasteless, inert.

Substances resembling Senega: Veratrum Viride, Arnica, Valerian, Serpentary. All have no keel.

Composition.—The active principle of senega is *saponin*, a colourless amorphous glucoside, $C_{32}H_{54}O_{18}$, decomposed by HCl into a sugar, and *sapogenin* ($C_{14}H_{22}O_2$). Saponin is closely allied to *digitonin*, one of the active principles of digitalis.

Preparations.

1. Infusum Senegæ.—1 in 20. *Dose*, 1 to 2 fl.oz.
2. Tinctura Senegæ.—1 in 8. *Dose*, ½ to 2 fl.dr.

ACTION AND USES.

1. IMMEDIATE LOCAL ACTION AND USES.

Externally.—Applied to the mucous membrane of the nose or throat, in the form of powder (snuff), senega is a powerful irritant, causing reflex hyperæmia, sneezing, cough, and mucous flow. These effects are not employed, but are a key to its remote local action. Solutions of saponin injected under the skin are violent local irritants and general depressants; the heart, vessels, central and peripheral nervous system, and muscles being dangerously affected.

Internally.—The action of senega on the stomach and intestines is moderately irritant, large doses causing epigastric heat, sickness, and diarrhœa; and medicinal doses **deranging digestion.** The absence of severe general symptoms indicates the difficulty of its absorption by the stomach.

2. ACTION IN THE BLOOD, AND SPECIFIC ACTION.

Saponin passes through the blood to the tissues. Senega diminishes the frequency of the heart, and probably affects the circulation much like digitalis, but in a manner which is more uncertain or at least still obscure.

3. REMOTE LOCAL ACTION AND USES.

Saponin appears to be excreted in part by the bronchial mucosa, which it stimulates thus remotely as it does when locally applied. The circulatory, muscular, and nutritive activity of the tubes is increased; the mucous secretion rendered more abundant and watery; and the efferent nerves stimulated, so that reflex cough is the result. The total action is said to be **expectorant,** the bronchial contents being expelled with greater force, and in greater volume, *i.e.* more readily and easily. Senega is in common use as a stimulant expectorant in the second stage of acute bronchitis, in chronic bronchitis, and in dilated bronchi, to liquefy and evacuate the contents of the tubes or cavities, and stimulate the " weak " surface of the mucous membrane. It is manifestly contra-indicated in acute bronchitis, phthisis, and when the digestion is feeble or deranged. Saponin is probably excreted in part by the skin and kidneys, both of which it slightly stimulates, increasing the volume of urine, and its most important solid constituents.

Krameriæ Radix—RHATANY ROOT.—The dried root of Krameria triandra. Imported from Peru.

Characters.—About an inch in diameter, branches numerous, long, brownish-red and rough externally, reddish-yellow internally, strongly astringent, tinging the saliva red.

Composition.—Rhatany root contains from 20 to 45 per

cent. of *rhatania-tannic acid*, $C_{54}H_{24}O_{21}$, a red amorphous substance, the watery solutions of which first colour chloride of iron green and then precipitate it, but are not precipitated by tartar emetic.

Incompatibles.—Alkalies, lime-water, salts of iron and lead, gelatine.

Preparations.

1. <u>Extractum Krameriæ.</u>—Aqueous. *Dose*, 5 to 20 gr.
2. <u>Infusum Krameriæ.</u>—1 in 20. *Dose*, 1 to 2 fl.oz.
3. <u>Pulvis Catechu Compositus.</u>—1 in 5.
4. <u>Tinctura Krameriæ.</u>—1 in 8. *Dose*, 1 to 2 fl.dr.

ACTION AND USES.

The preparations of rhatany possess the **properties of tannic acid**, and may be employed for the same purposes (see *Acidum Tannicum*, page 337), except that they are obviously of no use in poisoning by antimony. The drug is not extensively ordered.

SAPINDACEÆ.

Guarana. (*Not Officinal.*)—The seeds of Paullinia sorbilis, reduced to powder after roasting, and made into a stiff paste with water. From Brazil.

Characters.—Cylindrical rolls of dried paste. Brazilian cocoa.

Composition.—Guarana contains no less than five per cent. of *caffein*, $C_8H_{10}N_4O_2$, the alkaloid of the coffee and tea plants; united, as in these, with *tannic acid*, *starch*, and *gum*.

Dose.—15 to 60 gr. in powder, or as infusion.

ACTION AND USES.

The action of guarana closely resembles that **of strong tea** or coffee. It is chiefly used in sick headache (megrim). See *Caffein* (page 271).

ERYTHROXYLACEÆ.

Coca. (*Not Officinal.*)—The leaves of <u>Erythroxylon Coca.</u> From <u>South America.</u>

Characters.—Leaves two inches long, petiolate, oval, entire, pointed at the blunt apex, with a slight odour of tea, and a bitter aromatic taste.

Composition.—Coca leaves contain a yellowish-white crystalline bitter alkaloid, *cocaïn*, $C_{17}H_{21}NO_4$, which is converted by heat into a second alkaloid, *ecgonin*, $C_9H_{15}NO_3$, benzoic acid, and methylic alcohol.

Dose.—½ to 4 dr. of the leaves ; ⅛ to 1 gr. of cocaïn.

ACTION AND USES.

Coca is believed to possess stimulant, restorative, or even **nutritive** properties, enabling persons who chew the leaf to undergo great muscular exertion with little or no fatigue. In animals the alkaloid causes great muscular restlessness or excitement, and finally convulsions ; the whole brain, medulla, and cord being powerfully stimulated. The pupils are dilated ; respiration rises in frequency, is disturbed in rhythm, and finally ceases. The heart is greatly accelerated by paralysis of the vagus ; the blood pressure first rises and then falls. The muscles themselves remain unaffected. The amount of urea is said to be diminished, as if from diminished metabolism ; but coca does not prolong the life of starved animals.

This drug has been used to prevent muscular exhaustion ; in wasting attended with increased formation of urea (azoturia) ; in convalescence ; in mental exhaustion ; and in the opium habit. It has somewhat disappointed expectation.

LINACEÆ.

Lini Semina—LINSEED.—The seeds of Linum usitatissimum, Flax. Cultivated in Britain.

Characters.—Small, oval, pointed, flat, with acute edges, smooth shining brown externally, yellowish-white within, of a mucilaginous oily taste.

Composition.—The seeds of flax contain a quantity of *mucilage*, chiefly in the *testa* or coat, and from ¼ to ½ of their weight of the officinal *fixed oil*. This consists chiefly of the glyceride of *linoleïc acid*, which has a powerful affinity for oxygen, and thus becomes resinoid on exposure, constituting it a " drying oil." The linseed meal, obtained after expression of the oil, consists chiefly of mucilage, proteids, salts, a little oil, but neither starch nor sugar.

Preparations.

1. **Farina Lini.**—Linseed Meal. Linseed ground and deprived of oil, and the cakes powdered.

From Farina Lini is prepared :

> *a.* Cataplasma Lini.—Mix Linseed Meal 4, with boiling Water 10, and add Olive Oil ½, stirring constantly. Freshly bruised linseed is better than the meal and olive oil.
>
> Linseed meal is also used in preparing all Cataplasmata except Cataplasma Fermenti.

2. **Infusum Lini.**—"Linseed Tea," 1 in 30, with Liquorice. *Dose,* ad libitum.

3. **Oleum Lini.**—Brown. Made by expression without heat.

ACTION AND USES.

1. IMMEDIATE LOCAL ACTION AND USES.

Externally.—Linseed meal is used only as the cataplasma, which is the poultice universally employed to **convey heat and moisture** to parts, and thus affect the nerves, circulation, and nutrition generally. The oil may be applied to burns, either pure or mixed with an equal quantity of lime-water, constituting carron oil, a substitute for Linimentum Calcis. It may also be used as a laxative in the form of enema.

Internally.—Infusum Lini, or "linseed tea," is a familiar demulcent drink, containing a large quantity of mucilage, which coats the surface of the pharynx and fauces, and thus relieves troublesome throat cough, especially when it is combined with a little stimulant lemon.

2. ACTION IN THE BLOOD, SPECIFIC AND REMOTE LOCAL ACTION.

Linseed tea is supposed to have a specific and remote local effect as a demulcent on the bronchi and urinary passages, but this is probably referable to the warm water only. It is, perhaps, slightly diuretic as oil of linseed becomes oxydised in the system (as it does on exposure to air), and is excreted by the kidneys as a resinoid body which stimulates these organs.

MALVACEÆ.

Gossypium—COTTON WOOL.—The hairs of the seed of several species of Gossypium, carded.

Preparations.

Pyroxylin. Gun Cotton.—Made by immersing the wool in a mixture of sulphuric and nitric acids, washing, draining, and drying.

From Pyroxylin is prepared :

    <u>Collodium.</u>—Made by dissolving Pyroxylin, 1 ; in Ether, 36 ; and Rectified Spirit, 12.

From Collodium is prepared :

<u>Collodium Flexile.</u>—Collodium, 48 ; Canada Balsam, 2 ; and Castor Oil, 1.

Non-officinal Preparation of Gossypium.

Absorbent Cotton Wool. Cotton wool deprived of its oil by washing with an alkali.

ACTION AND USES.

The action and uses of cotton wool in medical and surgical practice are sufficiently familiar. Absorbent cotton is now much employed.

Pyroxylin is introduced for the purpose of making collodion.

Collodion, when painted on the skin or other exposed part, instantly dries by evaporation of the ether, forming a fine film. This film serves as a protective to thin, inflamed, broken, or incised surfaces, preventing bed-sores, arresting hæmorrhage (as in leech-bites), and closing fissures or punctures made with aspirateurs or trochars in paracentesis. The *flexible collodion* does not contract on drying or readily crack, and is a better form for most of the above purposes.

AURANTIACEÆ.

<u>Aurantii Cortex</u>—Bitter-Orange Peel.—The dried outer part of the rind of the bitter orange, <u>Citrus Bigaradia.</u> From the ripe fruit imported from the <u>south of Europe.</u>

Characters.—Thin, of a dark orange colour, nearly free from the white inner part of the rind, having an aromatic bitter taste and fragrant odour.

Composition.—Orange peel contains 1 to $2\frac{1}{2}$ per cent. of volatile oil, *oleum corticis aurantii*, isomeric with oil of turpentine, $C_{10}H_{16}$, and a bitter crystalline principle, *aurantiin* or *hesperidin*.

Preparations.

<u>1. Infusum Aurantii.</u>—1 in 20. *Dose*, 1 to 2 fl.oz.

<u>2. Infusum Aurantii Compositum.</u>—1 in 40. *Dose*, 1 to 2 fl.oz.

3. **Tinctura Aurantii.**—1 in 10.   *Dose,* 1 to 2 fl.dr.

   From Tinctura Aurantii is prepared:

   a. **Syrupus Aurantii.**—1 of tincture in 8.

   Tinctura Aurantii is also an ingredient of Mistura Ferri Aromatica and Tinctura Quiniæ.

4. **Vinum Aurantii.**—Orange wine made in Britain, and containing 12 per cent. of alcohol.

   Vinum Aurantii is used in making Vinum Ferri Citratis and Vinum Quiniæ.

   Bitter-orange peel is also an ingredient of Spiritus Armoraceæ Compositus, Tinctura Cinchonæ Composita, and Infusum Gentianæ Compositum, Mistura Gentianæ, and Tinctura Gentianæ Composita.

Aurantii Fructus—BITTER ORANGE.—The ripe fruit of Citrus Bigaradia.   Imported from the south of Europe.

Preparation.

Tinctura Aurantii Recentis.—6 in 20.   *Dose,* 1 to 2 fl.dr.

Aqua Aurantii Floris—ORANGE-FLOWER WATER. —Water distilled from the flowers of the bitter-orange tree, Citrus Bigaradia, and of the sweet-orange tree, Citrus aurantium.   Prepared mostly in France.

   Characters.—Nearly colourless, fragrant.
   Composition.—Orange flowers yield an aromatic volatile oil, *oleum neroli,* and a trace of a bitter principle.
   Impurities.—Lead derived from the vessels in which it is imported; detected by H_2S.   *Dose,* $\frac{1}{2}$ to 1 fl.oz.

Preparation.

Syrupus Aurantii Floris.—*Dose,* 1 to 2 fl.dr.

ACTION AND USES.

Orange is at once an **aromatic and a** bitter substance, and combines the action of these two classes of remedies, as described under *Calumba* and *Caryophyllum* respectively.   It is extensively used as a highly agreeable flavouring agent in cookery, pharmacy, and the manufacture of liqueurs; and in these several ways may be turned to account therapeutically. It is but feebly bitter.

Limonis Cortex—LEMON PEEL.—The outer part of the rind of the fresh fruit of Citrus Limonum. Lemons are imported from southern Europe.

Composition.—Lemon peel contains the officinal volatile oil, *Oleum Limonis,* $C_{10}H_{16}$ (isomeric with turpentine), and a bitter principle.

Preparations.

1. **Oleum Limonis.**—The oil expressed or distilled from the fresh peel. Pale yellow. *Dose,* 1 to 4 min.
   Oil of Lemon is an ingredient of Linimentum Potassii Iodidi cum Sapone and Spiritus Ammoniæ Aromaticus.
2. **Syrupus Limonis.**—2 in 41, with 20 of Lemon Juice. *Dose,* 1 to 2 fl.dr.
3. **Tinctura Limonis.**—1 in 8. *Dose,* ½ to 2 fl.dr.

Lemon peel is also contained in Infusum Aurantii Compositum and Infusum Gentianæ Compositum.

ACTION AND USES.

The action and uses of lemon are the same as those of orange, the only difference of importance being in the flavour. See *Aurantii Cortex.*

Limonis Succus—LEMON JUICE.—The freshly expressed juice of the ripe fruit of Citrus Limonum.

Characters.—A slightly turbid yellowish liquor, with a sharp acid taste, and grateful odour. Half a fluid ounce (one table-spoonful) contains 16·25 gr. of Citric Acid, and neutralises 23 gr. nearly of Bicarbonate of Potash, 20 gr. nearly of Bicarbonate of Soda, or 13 gr. fully of Carbonate of Ammonia.

Composition.—Lemon juice contains *citric acid*, both free and combined with *potash* and other bases, malic and phosphoric acids, etc.

Dose.—½ to 4 fl.oz.

Preparations.

1. **Syrupus Limonis.**—See *Limonis Cortex.*
2. **Acidum Citricum.**—See *Acids,* page 126.

o—8

ACTION AND USES.

1. IMMEDIATE LOCAL ACTION AND USES.

Lemon juice in the mouth and stomach has the same action as citric acid, and is used chiefly to **relieve thirst** and produce **effervescing** mixtures and drinks.

2. ACTION ON THE BLOOD, AND SPECIFIC ACTION AND USES.

Lemon juice enters the blood as alkaline citrates, potash salts, and phosphoric acid. Here the citrates are in part oxydised into carbonic acid and water. (See *Acidum Citricum.*) The potash and phosphoric acid probably act upon the red corpuscles, of which they are both important constituents.

Lemon juice is used with great success in the prevention and **treatment of scurvy,** a disease the exact nature of which is still obscure, but which is no doubt produced by the want of the juices of fresh vegetable and animal food. The citric acid, the potash, and the phosphoric acid have severally been credited with the beneficial effect by different authorities. Lemon juice has also been given in acute rheumatism, but is probably useful only in as far as it conveys alkalies into the blood and tissues.

3. REMOTE LOCAL ACTION AND USES.

These, which are of great importance, are fully described under Citric Acid.

Belæ Fructus—BAEL FRUIT.—The dried half-ripe fruit of Ægle Marmelos. From Malabar and Coromandel.

Characters.—Fruit roundish, about the size of a large orange, with a hard woody rind; usually imported in dried slices, or in fragments consisting of portions of the rind and adherent dried pulp and seeds. Rind about a line and a half thick, covered with a smooth pale-brown or greyish epidermis, and internally, as well as the dried pulp, brownish-orange or cherry-red. The moistened pulp is mucilaginous.

Composition.—Bael is believed to contain a kind of tannic acid, but has not been thoroughly analysed.

Preparation.

Extractum Belæ Liquidum.—1 in 1. *Dose,* 1 to 2 fl.dr.

ACTION AND USES.

In the fresh state, Indian bael is a pleasant refreshing fruit, with **astringent and** refrigerant properties, which render it valuable in the treatment of diarrhœa and dysentery. As imported into this country in hard portions of rind and dried pulp it is probably useless; but a liquid extract made from the fresh fruit appears to produce its specific effects. It is seldom employed out of India.

BYTTNERIACEÆ.

Oleum Theobromæ.—OIL OF THEOBROMA.—Synonym : Cacao Butter. A concrete oil obtained by expression and heat from the ground seeds of Theobroma cacao, a small tree, a native of Demerara and Mexico.

Characters.—Of the consistency of tallow; colour yellowish, odour resembling that of chocolate; taste bland and agreeable; fracture clean, presenting no appearance of foreign matter. Does not become rancid from exposure to the air. Melts at 95°. The seeds also contain theobromin. See *Caffein*, page 270.

Composition.—Oil of theobroma constitutes from 30 to 50 per cent. of the cacao bean. It consists chiefly of stearin with a little olein.

Preparations.

All the suppositories.

ACTION AND USES.

Cacao butter serves as a vehicle for more active substances in the form of suppositories. The action of theobromin is the same as that of caffein. See page 271.

CAMELLIACEÆ.

Tea. (*Not Officinal.*)—The dried leaves of Thea sinensis.

Composition.—Tea contains an alkaloid, *thein,* $C_8H_{10}N_4O_2$, identical with caffein; a *volatile oil*, most abundant in green tea, and *tannin*. The relations of the alkaloid, as well as its

ACTION AND USES,

are described fully under *Caffein*, page 271.

GUTTIFERÆ.

Cambogia—GAMBOGE.—A gum-resin obtained from Garcinia Morella. Imported from Siam.

Characters.—Cylindrical pieces, breaking easily with a smooth conchoidal glistening fracture; colour tawny, changing to yellow when it is rubbed with water; taste acrid.

Dose.—1 to 4 gr.

Impurity.—Starch; detected by yielding a green colour with iodine.

Composition.—Gamboge contains about 73 per cent. of a resinous substance, *gambogic acid*, $C_{20}H_{23}O_4$; 25 per cent. of *gum;* and about 2 per cent. of water. Gambogic acid is insoluble in water, gives the brilliant yellow colour to the gum-resin, and forms salts with bases. It is less active than the gum-resin.

Preparation.

Pilula Cambogiæ Composita.—Gamboge, 1; Barbadoes Aloes, 1; Compound Powder of Cinnamon, 1; Hard Soap, 2; Syrup, q. s. *Dose,* 5 to 10 gr.

ACTION AND USES.

1· IMMEDIATE LOCAL ACTION AND USES.

Gamboge is an **irritant** to the stomach and bowels, causing vomiting in large doses, and in medicinal doses acting as a **hydragogue** cathartic not unlike colocynth, without being cholagogue. It is seldom prescribed alone, and not often as the compound pill. Such a remedy is indicated in dropsies, cerebral hyperæmia, and as an anthelmintic (not to children); but other substances have now almost completely displaced it.

2. ACTION IN THE BLOOD, SPECIFIC AND REMOTE LOCAL ACTION AND USES.

Gambogic acid is chiefly thrown out in the liquid fæces; but part is absorbed, passes through the blood and tissues, and is excreted by the kidneys. These it stimulates, causing an increased flow of yellow-coloured urine. Its **diuretic** effect may add to its value in dropsy.

CANELLACEÆ.

Canellæ Albæ Cortex—CANELLA ALBA BARK. —The bark of Canella alba. From the West Indies.

Characters.—In quills or broken pieces, hard, of a yellowish-white or pale-orange colour, somewhat lighter on the internal surface. It has an aromatic clove-like odour, and an acrid peppery taste.

Composition.—Canella contains a *bitter principle* and an *aromatic oil.*

Canella Alba is contained in Vinum Rhei. (See *Rhei Radix,* page 318.)

ACTION AND USES.

Canella alba is an **aromatic** bitter stomachic and tonic, like cascarilla and cinnamon. The action and uses of this class of remedies is fully described under *Calumbæ Radix,* page 181.

VITACEÆ.

Uvæ—RAISINS.—The ripe fruit of <u>Vitis vinifera,</u> the Grape Vine, dried in the sun or with artificial heat. Imported from <u>Spain.</u>

Composition.—Raisins contain *grape sugar, acid tartrate of potash,* other vegetable acids, etc.

Raisins are contained in Tinctura Cardamomi Composita and Tinctura Sennæ.

ACTION AND USES.

Raisins are **demulcent,** refreshing and nutrient, but are employed in medicine chiefly as **sweetening** and flavouring agents.

ZYGOPHYLLACEÆ.

Guaiaci Lignum—GUAIACUM WOOD.—The wood of Guaiacum <u>officinale.</u> Imported from <u>St. Domingo</u> and <u>Jamaica,</u> and reduced by the turning lathe to the form of a coarse powder or small chips.

Guaiaci Lignum is an ingredient of Decoctum Sarsæ Compositum. (See *Sarsæ Radix,* page 354.)

Guaiaci Resina—GUAIACUM RESIN.—The resin of <u>Guaiacum officinale.</u> Obtained from the stem by natural exudation, by incisions, or by heat.

Characters.—In large masses of a brownish or greenish-brown colour; fractured surface resinous, translucent at the edges; with pleasant aromatic odour and burning taste. Insoluble in water; soluble in alcohol, ether, chloroform, and alkaline fluids. A solution in rectified spirit strikes a clear blue colour when applied to the inner surface of a paring of raw potato.

Substances resembling Guaiacum Resin: Myrrh, Scammony, Benzoin, Aloes, Resin, which have no green tinge.

Composition.—The chief constituent of guaiacum wood is the officinal *resin*, with a crystalline bitter colouring matter, gum, etc. The resin is itself composed of three resins, *guaiaconic acid*, $C_{19}H_{20}O_5$, 70 per cent.; *guaiac acid*, $C_6H_8O_3$, resembling benzoic acid; and *guaiaretic acid*, $C_{20}H_{26}O_4$, 10 per cent., with an indifferent resin.

Incompatibles.—Mineral acids, spirit of nitrous ether.

Dose.—10 to 30 gr.

Preparations.

1. **Mistura Guaiaci.**—Guaiacum Resin, 2; Sugar, 2; Gum Acacia, 1; Cinnamon water, 80. *Dose,* $\frac{1}{2}$ to 2 fl.oz.
2. **Tinctura Guaiaci Ammoniata.**—1 in 5 of Aromatic Spirit of Ammonia. *Dose,* $\frac{1}{2}$ to 1 fl.dr., with 1 drachm of mucilage or yolk of egg.
3. **Pilula Hydrargyri Subchloridi Composita.**—1 in $2\frac{1}{2}$. (See *Hydrargyrum*, page 86.)

ACTION AND USES.

1. IMMEDIATE LOCAL ACTION AND USES.

Internally, guaiacum is a local stimulant, producing salivation, an acrid hot sensation in the throat, warmth in the epigastrium, increase of the movements and secretions of the stomach and bowels, and reflex stimulation of the heart. In large quantity it is a gastro-intestinal irritant, causing powerful vomiting and purging, and the attendant disturbances of the system generally.

Guaiacum powder frequently relieves sore throat, if given in 30-grain doses, to be placed on the tongue, and slowly swallowed every six hours. The tincture or non-officinal lozenge is less successful. Plummer's pill doubtless owes part of its mildly purgative effect to the guaiac resin it contains.

2. ACTION ON THE BLOOD, SPECIFIC AND REMOTE LOCAL ACTION AND USES.

The further action of guaiacum physiologically is still obscure. Besides its stimulant effect on the circulation already mentioned, it appears to increase the secretions of the skin and kidney, and probably stimulates the liver and metabolism generally. In the form of the ammoniated tincture it is used in chronic gout and rheumatism, certainly with much success in some cases. As a constituent of Decoctum Sarsæ Compositum, not alone, it is given as an **alterative** in syphilis.

RUTACEÆ.

Buchu Folia.—BUCHU LEAVES.—The dried leaves of : 1. Barosma betulina, *Bartling*. 2. Barosma crenulata, *Hooker*. 3. Barosma serratifolia, *Willd*. Imported from the Cape of Good Hope.

Characters.—Smooth, marked with pellucid dots at the indentations and apex ; having a powerful odour and a warm camphoraceous taste. 1. About three-quarters of an inch long, coriaceous, obovate, with a recurved truncated apex and sharp cartilaginous spreading teeth. 2. About an inch long, oval-lanceolate, obtuse, minutely crenated, five-nerved. 3. From an inch to an inch and a half long, linear-lanceolate, tapering at each end, sharply and finely serrated, three-nerved.

Impurities.—Leaves of *Emplanum serrulatum* (for those of B. serratifolia) ; have no glands.

Substances resembling Buchu : Senna and Uva Ursi, which have entire leaves.

Composition.—Buchu contains a *volatile oil*, in the glands or "dots," of a yellowish-brown colour, and the source of the peculiar odour of the leaves ; a camphor, *barosma camphor ;* a crude oil ; and other substances of less importance.

Dose.—20 to 40 gr.

Preparations.

1. Infusum Buchu.—1 in 20.   *Dose,* 1 to 4 oz.
2. Tinctura Buchu.—1 in 8.   *Dose,* 1 to 4 fl.dr.

ACTION AND USES.

The action and uses of buchu closely resemble those of pareira, to the description of which the student is referred. It is more frequently employed than pareira, its infusion constituting an excellent vehicle for saline diuretics.

Oleum Rutæ.—OIL OF RUE.—The oil distilled from the fresh herb of Ruta graveolens.

Characters.—Colour pale yellow, odour disagreeable, taste bitter, acrid.

Composition.—Oil of rue is a mixture of various *volatile oils.*

Dose.—2 to 6 min.

ACTION AND USES.

The action of rue is the same as that of savin; but it is seldom employed as an **emmenagogue.** See *Sabinæ Cacumina,* page 351.

Cuspariæ Cortex—CUSPARIA BARK. ANGUSTURA BARK.—The bark of Galipea Cusparia. From tropical South America.

Characters.—In straight pieces, more or less incurved at the sides, from half a line to a line in thickness, pared away at the edges; epidermis mottled, brown, or yellowish-grey; inner surface yellowish-brown, flaky; breaks with a short fracture; the taste is bitter and slightly aromatic. The cut surface examined with a lens usually exhibits numerous white points or minute lines.

Impurity.—The bark of Strychnos Nux vomica ("false angustura bark"), which may be distinguished by its inner surface giving a blood-red colour with nitric acid, whilst true cusparia bark does not. Cusparia resembles Canella alba, but is darker and has pared edges.

Incompatibles.—Mineral acids; perchloride of iron, and other metallic salts.

Composition.—Cusparia contains a neutral crystalline bitter principle, *cusparin* or *angusturin,* a second bitter substance, an *aromatic oil,* but no tannin.

Dose.—10 to 40 gr.

Preparation.

Infusum Cuspariæ.—1 to 20.   *Dose,* 1 to 2 fl.oz.

ACTION AND USES.

Cusparia belongs to the group of **aromatic bitters,** the action and uses of which are fully discussed under *Calumba* and *Caryophyllum.* Like other bitters, it has been credited with antipyretic and antiperiodic properties, and in its native place is used instead of cinchona for malarious diseases.

Pilocarpi Folia *(Not Officinal).*—The leaflets

of <u>Pilocarpus pennatifolius.</u>   <u>Jaborandi.</u>   Imported from <u>Brazil.</u>

Characters.—Leaves dull green, large, pinnate, with 3 to 5 pairs of leaflets and a terminal one.   Leaflets coriaceous, 4 to 6 inches long, oblong, lanceolate, emarginate, smooth or only slightly tomentose, and full of pellucid dots.

Impurities.—Leaves of species of piper; not pinnate.

Composition.—Jaborandi contains *pilocarpin,* a liquid colourless alkaloid, to which its chief effects are due.   It is said to contain a second (isomeric) alkaloid, *jaborin,* closely resembling atropia in its action, and therefore antagonistic to pilocarpin.

Dose.—5 to 60 gr.

Non-officinal Preparations.

<u>Extract</u>, *Dose*, 2 to' 10 gr.   <u>Fluid Extract</u>, 1 in 1; *Dose*, 10 to 60 min.   <u>Infusion,</u> 1 in 20; *Dose.* 1 to 2 fl.oz.   And <u>Tincture,</u> 1 in 2; *Dose,* 5 to 20 min.   Also <u>Hydrochlorate and</u> <u>Nitrate</u> of Pilocarpin: *Dose of either,* $\frac{1}{20}$ to $\frac{1}{2}$ gr. by the mouth, $\frac{1}{10}$ to $\frac{1}{3}$ gr. hypodermically.   The last is most used.

ACTION AND USES.

1. IMMEDIATE LOCAL ACTION AND USES.

Externally.—Jaborandi applied to the conjunctiva causes contraction of the pupil, tension of the apparatus of accommodation and disturbance of vision.   The effect commences in ten minutes, and lasts from $1\frac{1}{2}$ to 24 hours before finally disappearing.   It is used in some cases of inflammation of the eye, such as iritis; in certain forms of blindness; and in paralysis of the muscles.   (See *Physostigma*, page 230.)

Internally, in full doses, it is liable to cause nausea and vomiting.

2. ACTION ON THE BLOOD.

Pilocarpin enters the blood rapidly and passes thence into the tissues.

3. SPECIFIC ACTION AND USES.

The striking effects of jaborandi consist in **profuse salivation, perspiration, disturbances of vision,** and **circulatory depression,** which last for hours, and leave a sense of drowsiness and debility behind them.   Salivation is due to stimulation of the terminal ends of the chorda tympani in the glands, as well as of its centre.   The flow commences in about five minutes after a moderate dose, and lasts several hours.   It increases with the dose.   It is completely prevented or arrested by atropia.

Perspiration is referable to stimulation both of the sudoriparous nerves and the sweat centres. It follows quickly on the appearance of the salivation; is accompanied by flushing of the skin, and sometimes rigor; progresses from the head downwards; may be so profuse as to soak the bedclothes; and lasts several hours. The body weight necessarily falls, metabolism is stimulated, and a large quantity of urea is said to be excreted by the skin. Atropia arrests this diaphoresis. The milk is doubtfully increased. The hair grows more actively under a course of jaborandi. Bronchial and nasal secretions flow more freely; even the tears, cerumen, and alimentary secretions are somewhat increased; but not the bile. The amount of urine is moderately raised by small doses. The menses are not affected. The eye is affected specifically, as it is locally. Respiration is not modified directly by pilocarpin. At first the heart and pulse are accelerated, but they are afterwards slowed and weakened; the blood pressure falls temporarily, then rises, and finally falls. Part of these effects are due to the action of the drug on the vagus in the heart, and can be arrested by atropia; part seem referable to the ganglia. The temperature rises before, and falls during, the sweating.

Pilocarpin has been tried in every kind of disease, but is now chiefly given as being a powerful and rapid diaphoretic. In renal dropsy, especially with uræmia, it may be of much service, eliminating a quantity of urea; also in effusions into the pleura and peritoneum; rarely in cardiac dropsy, since in this and every class of case it cannot be safely used if the heart be already weak. It has also been given in syphilis, and in a variety of uterine conditions, with various results. Bronchial catarrh, asthma, and pertussis are all relieved by the flux which it establishes. Small doses relieve the thirst of chronic Bright's disease. In certain dry skin diseases, and certainly in alopecia (baldness), it may answer well. Very conflicting reports have been published of its value in diphtheria, where it is said to loosen or detach the false membrane.

SIMARUBACEÆ.

Quassiæ Lignum—QUASSIA WOOD.—The wood of Picræna excelsa. From Jamaica.

Characters.—Billets varying in size, seldom thicker than the thigh. Wood dense, tough, yellowish white, intensely and purely bitter. Also chips of the same.

Substance resembling quassia : Sassafras, which is aromatic, and not bitter.

Composition.—The active principle of quassia is *quassin,* $C_{10}H_{12}O_3$, a white crystalline, neutral bitter principle. Quassia contains no tannin.

Preparations.

1. **Extractum Quassiæ.**—Aqueous. 48 in 1. *Dose,* 3 to 5 gr.
2. **Infusum Quassiæ.**—1 in 80 of *cold* water. *Dose,* 1 to 2 fl.oz.
3. **Tinctura Quassiæ.**—1 in 27. *Dose,* 1 to 2 fl.dr.

ACTION AND USES.

Quassia is a pure or simple bitter, and possesses the various properties fully described under *Calumbæ Radix.* It is very extensively used. The special points to be noted respecting it are : (1) that its preparations contain no tannin, and may be combined with salts of iron; (2) that it is entirely devoid of flavour, and intensely bitter, *i.e.* less agreeable than gentian and chiretta ; and (3) that the infusion is an excellent anthelmintic enema.

CELASTRACEÆ.

Euonymus—WAHOO. (*Not Officinal.*)—The bark of Euonymus atropurpureus.

Characters. — Quilled or curved pieces ; ash-grey with blackish patches without ; whitish within ; nearly inodorous ; sweetish, somewhat bitter, and acrid.

Composition. —Euonymus contains *euonymin,* an uncrystallisable intensely bitter principle, various *resins,* and a *fixed oil.*

Non-officinal Preparations.

Extractum Euonymi (U. S. P).—*Dose,* 1 to 5 gr.

Euonymin.—An eclectic preparation, consisting of the resins and fixed oil. *Dose,* 1 to 5 gr.

ACTION AND USES.

Euonymin is an hepatic stimulant, **direct cholagogue, and mild cathartic**; the latter effect being but little marked unless other purgatives be combined. It is used in constipation and hepatic derangements.

RHAMNACEÆ.

Rhamni Succus—BUCKTHORN JUICE.—The recently expressed juice of the ripe berries of common Buckthorn, Rhamnus catharticus, a native of Britain.

Characters.—Deep red by reflected, green by transmitted, light.

Composition.—Buckthorn juice contains an active principle, *rhamnicin*, probably identical with cathartic acid, the purgative glucoside of senna and rhubarb.

Preparation.

Syrupus Rhamni.—*Dose,* 1 fl.dr.

ACTION AND USES.

Buckthorn is an active cathartic, naturally resembling senna and rhubarb. It is very seldom employed.

Rhamnus Frangula. (*Not Officinal.*)—The bark of the Rhamnus frangula, or Black Alder. Imported from Holland.

Characters.—Quills half a line thick, with a warty greyish-brown exterior; nearly inodorous; taste sweetish bitter.

Composition.—Black Alder contains an active crystalline principle, *emodin*, also found in Rheum (*see* page 317).

Non-officinal Preparations.

A Fluid Extract, and Lozenges.
Dose.—1 to 4 fl.dr. of the former.

ACTION AND USES.

Rhamnus frangula is a certain and pleasant aperient, without griping or severe cathartic action, used in chronic constipation, and especially suitable for children.

Rhamnus Purshiana. (*Not Officinal.*)—The bark of Cascara sagrada, or Rhamnus purshiana. From the North Pacific coast.

Composition.—A crystalline and various resinoid bodies have been obtained from cascara, which requires to be further investigated.

Non-officinal Preparations.

A Fluid Extract, and a Cordial.

ACTION AND USES.

Cascara sagrada is a **tonic** and **stomachic** in small doses, **aperient** in large doses, and cathartic if freely given. It is useful in the same class of cases as the Rhamnus frangula.

ANACARDIACEÆ.

Mastiche—MASTICH.—A resinous exudation obtained by incision from the stem of Pistacia lentiscus. Produced in the island of Scio.

Characters.—Small irregular yellowish tears, brittle, becoming soft and ductile when chewed, having a faint agreeable odour.

Substances resembling mastich : Acacia, ammoniacum, galbanum, which are larger, rougher, and more opaque.

Composition.—Mastich consists of 80 or 90 per cent. of a resin, *mastichic acid,* soluble in alcohol; of a smaller quantity of another resin, *masticin,* soluble in ether, but insoluble in alcohol; and of a trace of volatile oil.

ACTION AND USES.

Mastich was formerly used much like other oleo-resins, but its application is now confined to dentistry, where it is employed as a **temporary stopping for carious** teeth. A solution in ether or collodion is applied on cotton wool with oil of cloves or cinnamon, and remains as a firm plug by evaporation of the solvent.

AMYRIDACEÆ.

Myrrha—MYRRH.—A gum-resinous exudation from the stem of Balsamodendron myrrha. Collected in Arabia Felix and Abyssinia.

Characters.—In irregular-shaped tears or masses varying much in size, somewhat translucent, of a reddish-yellow or reddish-brown colour; fractured surface irregular and somewhat oily; odour agreeable and aromatic; taste acrid and bitter.

Composition.—Myrrh contains about 2 per cent. of an oxygenated *æthereal oil,* $C_{10}H_{11}O$, *myrrhol;* a *resin, myrrhin,* 35 per cent.; and *gum* 60 per cent. Myrrh forms a milky-white emulsion with water, the resin being suspended by the gum in solution

Impurities.—Every variety of resins and gum-resins: detected by appearance, smell, and taste.

Preparations.

1. **Pilula Aloes et Myrrhæ.**—1 in 6.   (See *Aloe Socotrina*, page 357.)
2. **Tinctura Myrrhæ.**—1 in 8.   *Dose,* $\frac{1}{2}$ to 1 fl.dr.

Myrrh is also contained in Decoctum Aloes Compositum, Mistura Ferri Composita, Pilula Assafœtida Composita, and Pilula Rhei Composita.

ACTION AND USES.

1. IMMEDIATE LOCAL ACTION AND USES.

Externally.—Myrrh is a **stimulant and disinfectant** like other oleo-resins, and is sometimes used as a dressing for ulcers.

Internally.—It exerts a similar effect upon the mouth, throat, stomach, and bowels.   It is much employed as a wash in spongy gums and ulcerated mouth; as a gargle in relaxed throat; and as a **stomachic** and adjuvant of purgatives in dyspepsia, anæmia, and constipation.

2. ACTION ON THE BLOOD AND ITS USES.

Myrrh appears to **increase the number of leucocytes** in the blood; and this fact may in part account for its value along with iron in anæmia.

3. SPECIFIC ACTION.

Nothing definite is known on this subject.

4. REMOTE LOCAL ACTION AND USES.

Like the oleo-resins (see *Terebinthinæ Oleum*) myrrh appears to be excreted by the mucous membranes, especially of the genito-urinary and respiratory tracts, and stimulates them during its passage.   It is thus an **uterine stimulant and emmenagogue,** and is extensively given along with aloes or iron in the amenorrhœa of girls.   As a **stimulant and disinfectant expectorant** it is much less used now than formerly in chronic bronchitis.

Elemi—ELEMI.—A concrete resinous exudation, the botanical source of which is undetermined, but is probably Canarium commune.   Chiefly imported from Manilla.

Characters.—A soft unctuous adhesive mass, becoming harder and more resinous by age; of a yellowish-white colour;

with a rather fragrant fennel-like odour ; almost entirely soluble in rectified spirit.

Substances resembling Elemi : Assafœtida, Galbanum, Ammoniacum, known by smell.

Composition.—Elemi is a mixture of a turpentine and several resinous bodies.

Preparation.

Unguentum Elemi.—1 in 5.

ACTION AND USES.

Elemi acts much like resin of turpentine, and is employed in the ointment as a **stimulant and disinfectant** to sores and issues.

LEGUMINOSÆ.

Tragacantha—TRAGACANTH.—A gummy exudation from the stems of Astragalus verus, and possibly other species.   Collected in Asia Minor.

Characters. — White or yellowish, in broad shell-like slightly curved plates, tough and elastic, but rendered more pulverisable by a heat of 120° Fahr.   Very sparingly soluble in cold water ; but swelling into a gelatinous mass, which is tinged violet by tincture of iodine.   After maceration in cold water, the fluid portion is not precipitated by rectified spirit.

Impurities.—Other gums, and white lead.

Composition.—Tragacanth consists of two gums: *bassorin,* 33 per cent., comparatively insoluble in water, $C_{12}H_{20}O_{10}$, and unfermentable ; and a gum nearly identical with the *arabin* of acacia (but precipitated by acetate of lead), 53 per cent., soluble in water.   It also contains a little starch.

Preparations.

1. **Mucilago Tragacanthæ.**—1 in 80.   *Dose,* 1 fl.oz. or more.
2. **Pulvis Tragacanthæ Compositus.**—Tragacanth, 1 ; Gum Acacia, 1 ; Starch, 1 ; Sugar, 3.   *Dose,* 10 to 60 gr.

Tragacanth is also contained in Pulvis Opii Compositus.

ACTION AND USES.

1. IMMEDIATE LOCAL ACTION AND USES.

Internally, tragacanth is **demulcent.**   The mucilage may be used as a vehicle for active substances in linctuses for pharyngeal cough. Tragacanth is partly converted into sugar by the stomach; in large quantities it causes indigestion. It is chiefly employed to suspend resins and heavy powders, such as bismuth, the simple gum being preferable to the compound powder, because not fermentable.

2. ACTION ON THE BLOOD, SPECIFIC AND REMOTE LOCAL ACTION.

Tragacanth, like other gums, enters the blood and tissues, partly unchanged, partly as sugar and other products, and has a nutritive effect of comparatively low value. It is not used for this purpose. A remote demulcent effect on the urinary organs is probably imaginary only.

Glycyrrhizæ Radix—LIQUORICE ROOT.— The root or underground stem, fresh and dried, of Glycyrrhiza glabra. Cultivated in England.

Characters.—In long cylindrical branched pieces, an inch or less in diameter, tough and pliable; of a greyish-brown colour externally, yellow internally; without odour, of a sweet mucilaginous and slightly acrid taste. Digested with water, it yields a solution which gives a precipitate with diluted sulphuric acid.

Substances resembling Liquorice Root: Pyrethrum and Taraxacum, which are not sweet.

Composition.—Liquorice root contains *grape-sugar*, *glycyrrhizin, starch, resin, asparagin,* and *malic acid.* Glycyrrhizin is a yellow amorphous glucoside, $C_{24}H_{36}O_9$, with a strong bittersweet taste and acid reaction, yielding glucose and a very bitter substance, *glycyrretin.*

Preparations.

1. **Extractum Glycyrrhizæ.**—Aqueous. *Dose,* $\frac{1}{2}$ to 1 dr.
2. **Extractum Glycyrrhizæ Liquidum.**—Made as above with spirit. 2 fl. oz. = 1 oz. of solid extract. *Dose,* 1 fl. dr.
3. **Pulvis Glycyrrhizæ Compositus,**—1, with 1 of Senna and 3 of Sugar. *Dose,* 30 to 60 gr.

Non-officinal Preparation.

Pulvis Liquiritiæ Compositus (Ph. Germ. and Russ.).—1, with 1 of Senna, $\frac{1}{2}$ of Sulphur, $\frac{1}{2}$ of Fennel, and 3 of Sugar. *Dose,* a teaspoonful.

Liquorice or its preparations are contained in many preparations throughout the Pharmacopœia. It especially covers the taste of senna, chloride of ammonium, senega, hyoscyamus, turpentine, and bitter sulphates. The powdered root is a useful basis for pills.

ACTION AND USES.

Liquorice is chiefly used for the pharmaceutical purposes just indicated. It has a pleasant taste and flavour, and increases the flow of saliva and mucus when slowly chewed or sucked, the increased secretions acting as emollients to the throat. Liquorice is therefore a popular **demulcent**, much used to relieve sore throat and coughs.

Scoparii Cacumina.—Broom Tops.—The fresh and dried tops of Sarothamnus Scoparius. From indigenous plants.

Characters.—Straight angular dark-green smooth tough twigs, of a bitter nauseous taste, and of a peculiar odour when bruised.

Composition.—Scoparium contains two active principles, *scoparin* and *spartein*, besides other constituents. Scoparin $C_{21}H_{22}O_{10}$, is a yellow crystalline neutral body, said by some to be a diuretic, by others not so. Spartein, $C_{15}H_{26}N_2$, is a volatile oily-looking liquid alkaloid, allied in appearance, composition, and physiological action to conia. See *Conii Fructus*, page 250.

Preparations.

Decoctum Scoparii.—1 *dried* in 20. *Dose*, 2 to 4 fl.oz.

Succus Scoparii.—3 of juice of *fresh* tops to 1 of spirit. *Dose*, 1 to 2 fl.dr.

ACTION AND USES.

Broom has a bitter stomachic and somewhat **astringent** action in moderate doses, but is not used on this account. Its further effect on the system is still obscure, the only fact definitely known being that it frequently produces free diuresis. It is believed that the active principles of the plant, either or both, pass through the blood and tissues, and stimulate the secreting substance of the kidneys during the process of excretion. Broom is therefore extensively used in this country as a diuretic in dropsy, especially cardiac dropsy, but is almost invariably combined with other drugs of the same class, such as digitalis, acetate of potash, etc. It should be avoided in acute renal dropsy.

Pterocarpi Lignum.—Red Sandal Wood.—The wood of Pterocarpus santalinus. From Ceylon.

P—8

Characters.—Dense heavy billets, outwardly dark brown, internally variegated with dark and lighter red rings, if cut transversely. Powder blood-red, of a faint peculiar odour, and an obscurely astringent taste. Also chips of the same.

Substances resembling Sandal Wood : Logwood, less dense.

Composition.—Red sandal wood contains a blood-red crystalline principle, *santalic acid,* or *santalin,* insoluble in water.

Red Sandal Wood is contained in Tinctura Lavandulæ Composita.

USE.

Red sandal wood is used only to **give colour** to the Compound Tincture of Lavender.

Kino—Kino.—The inspissated juice obtained from incisions made in the trunk of Pterocarpus Marsupium. Imported from Malabar.

Characters.—In small angular brittle glistening reddish-black fragments, translucent and ruby-red on the edges, inodorous, very astringent. When chewed it tinges the saliva blood-red.

Composition.—Kino contains 75 per cent. of *kino-tannic acid* $C_{18}H_{18}O_8$, giving a greenish precipitate with persalts of iron; *brenzcatechin,* a derivate of catechin (see *Catechu Pallidum,* page 270); and *kino-red,* formed from kino-tannic acid by oxydation.

Dose.—10 to 30 gr.

Preparations.

1. **Pulvis Kino Compositus.**—Kino, 15; opium, 1; cinnamon, 4 (1 of opium in 20). *Dose,* 5 to 20 gr.
2. **Tinctura Kino.**—1 to 10. *Dose,* ½ to 2 fl.dr.

Kino is also a constituent of Pulvis Catechu Compositus, 1 in 5.

ACTION AND USES.

Kino closely resembles tannic acid in its action, and may be used for the same purposes. (See page 337.) It is chiefly employed in the form of **astringent** gargles, and as a constituent of mixtures for diarrhœa.

Balsamum Peruvianum—Balsam of Peru. —A balsam obtained from Myroxylon Pereiræ. It exudes from the trunk of the tree after the bark has been scorched and removed. From Salvador in Central America.

Characters.—A reddish-brown or nearly black liquid, translucent in thin films; having the consistence of syrup, a balsamic odour, and an acrid slightly bitter taste; soluble in five parts of rectified spirit. Undergoes no diminution in volume when mixed with water.

Impurity.—Resin, soluble in bisulphide of carbon.

Composition.—Balsam of Peru is a complex substance. The greater part consist of (1) the *volatile oil of Peruvian balsam,* which is itself composed of *cinnamin* (or *cinnamate of benzyl-æther*), $C_{16}H_{14}O_2$; *styracin* (or *cinnamate of cinnamic-æther*), $C_{18}H_{16}O_2$; *peruvin* (or benzyl-alcohol), C_7H_8O; *benzoate of benzyl æther;* (2) *cinnamic* and *benzoic acids* in small quantities; and (3) a mixture of *resins,* probably hydrates of cinnamin. See *Styrax.*

Dose.—10 to 15 minims, made into an emulsion with mucilage or yolk of egg.

ACTION AND USES.

1. IMMEDIATE LOCAL ACTION AND USES.

Externally.—Balsam of Peru possesses the properties of its several constituents—benzoic acid and its allies and resins, being an **antiseptic and disinfectant, a vascular and nutritive stimulant, and a nervine sedative.** (See *Terebinthinæ Oleum* for a full account, page 344.) Balsams have been used from time immemorial as applications to wounds and sores, but are now almost entirely displaced by simpler dressings, such as carbolic acid and boracic acid. They are still used, however, to cleanse bed-sores. A more important application of Peruvian balsam is in certain diseases of the skin, namely, (1) in some chronic inflammatory affections (eczema); (2) to **relieve itching,** prurigo, urticaria, etc., 1 in 8 of vaselin; (3) in scabies, for which it is the best of all remedies, **killing the acarus,** relieving the itching and inflammation, and disinfecting the parts. The skin should be thoroughly rubbed with it (1 drachm for the whole body) on two or more occasions, a warm bath being taken before and after.

Internally.—Balsam of Peru has a mild carminative effect on the stomach and bowels, like volatile oils.

2. ACTION ON THE BLOOD, SPECIFIC AND REMOTE LOCAL ACTION AND USES.

The important changes undergone in the blood and tissues by benzoic and cinnamic acids, and the excretion of these and of aromatic oils by the mucous membranes, kidneys, and skin, are fully discussed under *Benzoin, Styrax,* and *Terebinthinæ Oleum.* The constituents of Peruvian balsam appear chiefly to affect the respiratory organs; and it may therefore be added to cough mixtures as an agreeable **stimulant and disinfectant expectorant** in chronic bronchitis.

Balsamum Tolutanum—BALSAM OF TOLU.—

A balsam obtained from Myroxylon Toluifera. It exudes from the trunk of the tree after incisions have been made into the bark. From New Granada.

Characters.—A reddish-yellow soft and tenacious solid, becoming hard by keeping, with a fragrant balsamic odour; soluble in rectified spirit.

Composition.—Balsam of Tolu contains a *turpentine* $C_{10}H_{16}$, benzoic and cinnamic acids, and various *resins.*

Dose.—10 to 20 gr., as an emulsion with mucilage and sugar.

Preparations.

1. **Syrupus Tolutanus.**—1 in 29.   *Dose,* 1 to 2 fl. dr.
2. **Tinctura Tolutana.**—1 in 8.   *Dose,* 15 to 30 min.

Balsam of Tolu is also a constituent of Tinctura Benzoini Composita.

ACTION AND USES.

These are the **same as** those **of Peruvian balsam,** but tolu is used internally only, and chiefly as a pleasant ingredient of cough mixtures.

Physostigmatis Faba—CALABAR BEAN.—The

seed of Physostigma venenosum.   Western Africa.

Characters.—About the size of a very large horse-bean, with a very firm, hard, brittle shining integument of a brownish-red, pale chocolate, or ash-grey colour. Irregularly kidney-shaped, with two flat sides, and a furrow running longitudinally along its convex margin, ending in an aperture near one end of the seed. Within the shell is a kernel consisting of two cotyledons, weighing on an average about 46 grains, hard, white, and pulverisable, of a taste like that of the ordinary edible leguminous seeds, without bitterness, acrimony, or aromatic flavour. It yields its virtues to alcohol, and imperfectly to water.

Composition.—Besides the ordinary constituents of beans, the seed of physostigma contains an active principle, *physostigmin* or *eserin*, $C_{15}H_{21}N_3O_2$, an alkaloid, combining with acids, and variously obtained as colourless crystals, or an amorphous or syrupy body.

Dose, in powder.—1 to 4 gr.

Preparation.

Extractum Physostigmatis.—Spirituous. 45 in 1. *Dose,* $\frac{1}{16}$ to $\frac{1}{4}$ gr.

---

1. IMMEDIATE LOCAL ACTION AND USES.

Extract of physostigma or preparations of eserine are readily absorbed by the conjunctiva, and produce the specific contraction of the pupil to be presently noticed.

Taken by the mouth, calabar bean in moderate doses sometimes causes sickness and colic, and in larger doses diarrhoea, all from increased and irregular peristalsis, apparently of local origin. The extract is therefore occasionally used in habitual constipation.

2. ACTION IN THE BLOOD.

Eserin enters the blood unchanged, and passes thence into the tissues.

3. SPECIFIC ACTION AND USES.

Eserin is found in all the organs. Along with the gastro-intestinal symptoms first described, moderate doses of the bean give rise to a sense of weakness, faintness, and shortness of breath ; larger doses to an aggravation of the same symptoms, with contraction of the pupil, frontal headache, salivation, diaphoresis, slowing and weakening of the pulse. These are short of truly poisonous effects.

On analysis it is found that consciousness is not lost, though impaired by large doses, showing comparative freedom of the *convolutions.* The *cord* is the part principally affected by calabar bean, the chief symptoms being of the nature of motor paralysis from **depression of the anterior cornua,** and thus of reflex irritability also. The respiratory muscles necessarily fail from this cause. The posterior cornua (sensory portions) of the cord are paralysed to a degree, so that sensibility is diminished in the limbs. The *motor nerves* and *muscles* are but slightly affected directly. Occasional twitchings occur, whether direct or spinal in origin. The *sensory nerves* are not directly influenced. The *medulla* is decidedly affected by physostigma. Thus the **respiratory centre,** after brief (probably reflex) stimulation, is **depressed,** and death occurs chiefly by asphyxia. The cardiac centre is first stimulated, so that the heart beats more powerfully and less frequently; but at last, or after large doses, depression ensues. Therewith the intracardiac branches of the vagus are probably stimulated at first, and the ganglia paralysed at last. The blood pressure rises with the increased cardiac action, and falls later on. Whether there is any direct action of eserin on the vaso-motor apparatus is unsettled.

Contraction of the pupil and spasm of accommodation are striking and highly important effects of eserin, whether it be given internally or applied locally. Both phenomena are due to **irritation of the fibres of the third nerve,** and not to central disturbance as in the contraction caused by opium, or to paralysis of the sympathetic. These effects are accompanied by fall of the intraocular tension, and can be removed by atropia. The salivary secretion is increased through the centre of the chorda, but ceases after large doses from arrest of the circulation in the glands.

The specific *uses* of calabar bean depend on its action on the cord and the eye. It has been frequently given in tetanus, and other convulsive diseases referable to irritation or disease of the spinal centres, and apparently with success, although many of these cases recover spontaneously, and others resist the eserin. The alkaloid should be given subcutaneously in doses of gr. $\frac{1}{60}$ to $\frac{1}{12}$ in solution; or gr. $\frac{1}{3}$ of the extract may be given subcutaneously, or gr. 1 by the mouth, repeated in two hours, and followed by doses of gr. $\frac{1}{16}$ to $\frac{1}{4}$ every few hours. For the convulsions of strychnia poisoning calabar bean is of little or no use. Neither is it of much real service in the treatment of poisoning by atropia or chloral, as was once expected.

In diseases of the eye eserin is now much used. A drop of a solution of the sulphate (2 gr. to 1 ounce of water) is applied locally to diminish intraocular pressure in glaucoma, perforating keratitis, etc.; in paralysis of the iris and ciliary muscle, *e.g.* after diphtheria ($\frac{1}{2}$ gr. to 1 ounce); to counteract the effects of belladonna; or to diminish the entrance of light in painful diseases of the eye, photophobia, etc.

4. REMOTE LOCAL ACTION.

Eserin is excreted by the liver and salivary glands, but has never been found in the urine

Araroba — GOA POWDER. — A concretion from clefts in the stem of Andira Araroba (Angeliun Amargosa) imported from Brazil chiefly into India. (*Not officinal.*)

Characters.—A powder or in small pieces of a light yellow colour, becoming pale-brown by exposure; mixed with small pieces of wood.

Composition.—Goa powder contains 80 per cent. of *chrysarobin*, $C_{30}H_{26}O_7$, which is converted into *chrysophanic acid*, by the process presently to be given.

From Goa Powder is prepared:

Acidum Chrysophanicum.—Chrysophanic acid $C_{41}H_{10}O_4$. Made by exhausting Goa powder with hot Benzol, filtering, allowing Chrysarobin to crystallise out, dissolving this in strong Potash, and decomposing with a mineral acid. It may also be obtained from Rhubarb. See *Rhei Radix*, page 318.

Characters.—A dull orange-yellow powder, or shining yellow needles, odourless, with an acrid taste; insoluble in water, and spirit; soluble in hot benzol, oils, and vaseline.

Dose.—$\frac{1}{10}$ to 2 gr.

Preparation.

Unguentum Acidi Chrysophanici.—*Unguentum Chrysarobini* (U.S.P.).   1 in 10 of benzoated lard.

ACTION AND USES.

1. IMMEDIATE LOCAL ACTION AND USES.

Externally.—Goa powder or chrysophanic acid **destroys low vegetable organisms** in connection with the **skin**, stains it yellow, and **stimulates** it so much as to produce in some instances serious constitutional disturbance. It is a successful application in some forms of ringworm, and in scaly diseases of the skin, especially psoriasis.

Internally.—Chrysophanic acid is apt to cause vomiting and purging.

2. ACTION ON THE BLOOD; SPECIFIC ACTION; AND REMOTE LOCAL ACTION AND USES.

Chrysophanic acid has been given with various degrees of success in psoriasis, apparently by a remote local action on the skin.

Senna Alexandrina—ALEXANDRIAN SENNA.—The leaflets of Cassia lanceolata and Cassia obovata, imported from Alexandria; carefully freed from the flowers, pods, and leafstocks of the same, and from the leaves, flowers, and fruit of Solenostemma Argel.

Characters and Tests.—Lanceolate or obovate leaflets, about an inch long, unequally oblique at the base, brittle, greyish-green, of a faint peculiar odour, and mucilaginous sweetish nauseous taste. The unequally oblique base, and freedom from bitterness, distinguish the senna from the argel leaves, which, moreover, are thicker and stiffer.

Impurities; and *substances resembling Senna:* Soleno-stemma Argel, Uva Ursi, and Barosma, all equal at the base.

Senna Indica—Tinnivelly Senna.—The leaf-lets of Cassia elongata. From plants cultivated in Southern India.

Characters.—About two inches long, lanceolate, acute, unequally oblique at the base, flexible, entire, green, without any admixture; odour and taste those of Alexandrian Senna.

Composition.—Senna contains an active principle, *cathartic acid;* a colouring matter closely allied to *chrysophanic acid;* peculiar unfermentable sugar, *catharto mannite;* other obscure glucosides, *sennapicrin* and *sennacrol;* and various vegetable salts. Cathartic acid, a highly important body, is an amorphous glucoside, $C_{180}H_{192}N_2SO_{82}$, which forms salts with bases, and can be broken up in into glucose and cathartogenic acid.

Dose.—10 to 30 gr. in powder.

Preparations of either kind of Senna:

1. **Confectio Sennæ.**—1 in 11 with Coriander, Figs, Tamarinds, Cassia Pulp, Prunes, Extract of Liquorice, Sugar and Water. *Dose,* 60 to 120 gr.

2. **Infusum Sennæ.**—1 in 10. *Dose,* 1 to 2 fl.oz.

From Infusum Sennæ is prepared:

   a. **Mistura Sennæ Composita.**—Infusion of Senna, 14; Tincture of Senna, 2½; Sulphate of Magnesia, 4; Extract of Liquorice, ½; Compound Tincture of Cardamoms, 1¼. *Dose,* 1 to 1½ fl.oz.

3. **Syrupus Sennæ.**—1 in 2. *Dose,* 1 to 4 fl.dr.

4. **Tinctura Sennæ.**—1 in 8. *Dose,* 1 to 4 fl.dr.

Senna is also the most important ingredient in Pulvis Glycyr-rhizæ Compositum. 2 in 10. See *Glycyrrhizæ Radix,* page 224.

ACTION AND USES.

1. IMMEDIATE LOCAL ACTION AND USES.

Given *internally*, senna stimulates the muscular coat of the intestine, apparently by local reflex action, originating in the mucous surface of the bowel itself; and produces brisk peristaltic movements and purgation within four or five hours. The colon is chiefly stimulated, hurrying downwards the fluid contents received from the ileum, which appear as very thin copious yellow stools, with excess of soda salts and digestive products, but no special increase of bile. Full doses cause repeated evacuation and griping, but no inflammation of the mucous surface. The pelvic structures may, however, become hyperæmic, leading to hæmorrhoids and the appearance of the menses. Constipation does not follow the use of senna.

Senna is never given alone, but always with a carminative to prevent griping, and frequently with other purgatives, as in the compound mixture. It is one of the most useful of purgatives. It is very extensively prescribed to complete the effect of mercurial and other duodenal purgatives, given several hours before. It affords at once a rapid and a safe purge at the commencement of febrile attacks in children, in local inflammations, and in cerebral congestion. As an habitual laxative in the form of Pulvis Glycyrrhizæ Compositus, senna is most valuable as a simple stimulant of the muscular coat, which neither loses its effect by use, nor produces subsequent constipation. Combined with bitter and other stomachics, it is useful in atonic dyspepsia, its laxative effect being increased by acids but diminished by alkalies.

ACTION IN THE BLOOD, SPECIFIC AND REMOTE LOCAL ACTION.

Cathartic acid and chrysophanic acid enter the blood, pass through the tissues, and are excreted by the kidneys and mammary gland; the cathartic acid purging infants at the breast, the chrysophanic acid staining the urine yellow. Senna purges animals when injected into the veins.

Hæmatoxyli Lignum—LOGWOOD.—The sliced heart-wood of Hæmatoxylum Campechianum. Imported from Campeachy, Honduras, and Jamaica.

Characters.—The logs are externally of a dark colour, internally they are reddish-brown. The chips have a feeble agreeable odour, and a sweetish taste. A small portion chewed imparts to the saliva a dark pink colour.

Composition.—Logwood contains *tannic acid,* and a peculiar colouring principle, *hæmatoxylin,* $C_{10}H_{14}O_6$, occurring in colourless crystals, which become red on exposure to light, the solutions undergoing various changes of colour with acids and alkalies, and coagulating gelatine. The decoction precipitates perchloride of iron violet blue, acetate of lead and other metallic salts a beautiful blue. Other less important substances occur in logwood.

Incompatibles.—Mineral acids, metallic salts, lime-water, and tartar emetic.

Preparations.

1. **Decoctum Hæmatoxyli.**—1 in 20. *Dose,* 1 to 2 fl.oz.
2. **Extractum Hæmatoxyli.**—Aqueous. *Dose,* 10 to 30 gr.

ACTION AND USES.

Hæmatoxylum possesses the **astringent** action of tannic acid, and may be used in the same class of cases. See *Galla.*

Cassiæ Pulpa—CASSIA PULP.—The pulp obtained from the pods of the Purging Cassia, Cassia Fistula. Imported from the East Indies, or recently extracted from pods imported from the East or West Indies.

Characters of the pods.—Cylindrical, a foot or more in length, slightly curved, woody, indehiscent, black, rounded, divided by septa into cells, each containing a seed and viscid pulp. *Of the pulp :* blackish-brown, viscid, sweetish disagreeable taste, and somewhat sickly in odour, usually containing the seeds and dissepiments.

Composition.—Cassia pulp contains sugar, pectin, mucilage, and a purgative principle supposed to be allied to cathartic acid. See *Senna.*

Cassia Pulp is contained in Confectio Sennæ, about 1 in 8.

ACTION AND USES.

Cassia pulp is a **laxative,** given only as an ingredient of Confectio Sennæ.

Tamarindus—TAMARIND.—The preserved pulp of the fruit of Tamarindus indica. Imported from the West Indies.

Characters and Test.—A brown sweetish subacid pulp preserved in sugar, containing strong fibres and brown shining seeds, each enclosed in a membranous coat.

Impurity.—Copper; a piece of bright iron left in contact with the pulp for an hour does not exhibit any deposit of copper.

Composition.—Tamarind contains sugar, gum, tartaric acid, acid tartrate of potash, citric, acetic, and various aromatic acids.

Tamarind is contained in Confectio Sennæ, about 1 in 8.

ACTION AND USES.

Tamarind acts as a pleasant acid **refrigerant and gentle laxative.** For the former purpose it is prepared either as an infusion or as tamarind whey (1 part of the pulp to 30 parts warm milk), which is also a mild purgative, like the Confectio Sennæ.

Copaiba—COPAIVA.—The oleo-resin obtained from incisions made in the trunk of Copaifera multijuga, and other species of Copaifera. Chiefly from the valley of the Amazon.

Characters and Tests.—About the consistence of olive oil, light yellow, transparent, with a peculiar odour, and an acrid aromatic nauseous taste. Perfectly soluble in an equal volume of benzol. Is not fluorescent.

Composition.—Copaiba consists of less than 50 per cent of the officinal *volatile oil*, and more than 50 per cent of *resin*. The oil of copaiva, isomeric with turpentine, $C_{10}H_{16}$, is colourless or pale yellow, with the odour and taste of copaiva. Resin of Copaiva, $C_{20}H_{30}O_2$, is a brownish resinous mass, consisting of a crystallisable resin, *copaivic acid*, the chief constituent of the oleo-resin, and a non-crystallisable *viscid resin of copaiba*, amounting to 1½ per cent. The proportion of oil and resin varies much with the age and exposure of the copaiba.

Impurities.—Turpentine, detected by the odour on heating. Fixed oils, detected by a greasy ring round the resinous stain left by copaiva when heated on paper. Copaiva dissolves one-fourth its weight of carbonate of magnesia by the aid of heat, and remains transparent; not so the fixed oil. Gurjun balsam, which coagulates at 270°.

Dose.—½ to 1 fl.dr.

Preparation.

Oleum Copaibæ.—The oil distilled from Copaiba. *Dose,* 5 to 20 min., with mucilage or yolk of egg.

ACTION AND USES.

1. IMMEDIATE LOCAL ACTION.

Copaiva produces an acrid nauseous sensation in the mouth, warmth in the stomach, unpleasant eructations, and gastro-intestinal irritation like other oleo-resins. Large doses or the persistent use of the drug leads to dyspepsia, sickness, and diarrhœa; and it is contra-indicated in irritable states of the stomach and bowels.

2. ACTION IN THE BLOOD, AND SPECIFIC ACTION.

The active principles of copaiva are absorbed into the blood, and pass thence into the tissues. The action of copaiva on the organs and tissues generally is obscure.

3. REMOTE LOCAL ACTION AND USES.

The volatile oil of copaiba is excreted by the kidneys, bronchi, and skin, and the resin at least by the kidneys. All the secretions smell freely of the drug, and the neighbourhood of the patient is pervaded with a characteristic unpleasant odour. In thus passing through the eliminating organs, copaiva stimulates them, altering their secretions and the nutrition of their cells and vessels. The urine is passed more frequently. and usually in increased quantity; but it may be scanty, with albumen and blood, pain in the loins, and other symptoms of renal congestion. The albumen thus passed must be distinguished from the acid resin of copaiva which may be thrown down from the urine by nitric acid, and which is dissolved by heat or alcohol. Carried by the urine into the bladder and urethra, and possibly also excreted by the mucous membranes of the same parts, copaiva produces along the whole **genito-urinary tract a stimulant and disinfectant effect.** A similar influence is produced in the bronchi, and the mucous secretion is increased, and **expectoration reflexly excited.** The stimulation of the skin (and probably the primary gastro-intestinal irritation in part) may sometimes cause an eruption, the " copaiba rash," not unlike that of measles.

The uses of copaiva depend entirely on its remote local effects, the immediate local effects only suggesting care in its administration. Its chief application is to the genito-urinary organs. The resin is given as a highly useful **diuretic in** hepatic and cardiac dropsy, but must be avoided in the dropsy attending Bright's disease. It is much to be preferred to the oleo-resin for this purpose. The latter is chiefly employed in inflammatory affections of the bladder and urethra, especially gonorrhœa, when the first acute symptoms have somewhat subsided. It is best combined with potash and cubebs. Naturally it is less useful in the vaginal gonorrhœa of women. Copaiba is now seldom used in bronchial affections, on account of the unpleasant effects attending it; but in hospital practice it will sometimes diminish and disinfect the profuse foul products of chronic bronchitis and bronchiectasis when other means have failed. It is occasionally given in skin diseases.

Acaciæ Gummi—Gum Acacia.—A . gummy exudation from the stems of one or more undetermined species of Acacia.

Characters and Tests.—In spheroidal tears, nearly colourless, and opaque from numerous minute cracks; or in fragments with shining surfaces; brittle; bland and mucilaginous in taste; insoluble in alcohol, but soluble in water.

Impurities.—Starch; detected by the iodine test. Gum resins; detected by smell and taste.

Incompatibles.—Alcohol and sulphuric acid, borax, persalts of iron, and subacetate of lead.

Composition.—Gum arabic consists chiefly of *arabic acid* or *arabin*, $C_{36}H_{66}O_{33}$, combined with calcium, magnesia, and potash, and 17 per cent. of water.

Preparations.

Mucilago Acaciæ.—Gum, 40; Water, 60. *Dose,* 1 to 4 fl.dr.

Gum Acacia is also contained in Mistura Cretæ, Mistura Guaiaci, Pulvis Amygdalæ Compositus, Pulvis Tragacanthæ Compositus, and in all Trochisci.

ACTION AND USES.

Acacia possesses very similar properties and physiological effects to those of tragacanth, and is employed for the same purposes. (See *Tragacantha.*) An objection to its pharmaceutical use is its hability to undergo fermentation, and cause indigestion, flatulence, and diarrhœa. Its principal application therapeutically is for cough in the form of lozenges and linctuses.

Indigo—C_8H_5NO. A blue pigment prepared from various species of Indigofera. From India.

Use.—Indigo is employed in chemical testing.

ROSACEÆ.

Rosæ Gallicæ Petala—RED ROSE PETALS.—The fresh and dried unexpanded petals of Rosa gallica. From plants cultivated in Britain.

Characters.—Colour fine purplish-red, retained after drying; taste bitterish, feebly acid, and astringent; odour roseate, developed by drying.

Composition. — Red-rose petals contain an *aromatic oil, tannic* and *gallic acids*, gum, colouring matters, salts, etc. *Oleum rosæ* exists in very small quantity; it consists of an aromatic oxygenated elœoptin, and an odourless solid rose-camphor.

Preparations.

1. Confectio Rosæ Gallicæ.—1 of *fresh* petals in 4. *Dose,* 30 to 60 gr.
2. Infusum Rosæ Acidum.—1 of *dried* petals in 40 of diluted sulphuric acid and water. *Dose,* 1 to 2 fl.oz.
3. Syrupus Rosæ.—1 of *dried* petals in 17¼. *Dose,* 1 to 2 fl.dr.

Rosæ Centifoliæ Petala — CABBAGE ROSE PETALS.—The fresh petals, fully expanded, of Rosa centifolia. From plants cultivated in Britain.

Characters.—Taste sweetish, bitter, and faintly astringent; odour roseate; both readily imparted to water.

Preparation.

Aqua Rosæ.—1 in 1 by distillation. *Dose,* 1 to 2 fl.oz.

ACTION AND USES.

The preparations of the red and the cabbage rose are chiefly used as pleasant vehicles. The acid infusion is an agreeable astringent.

Rosæ Caninæ Fructus—FRUIT OF THE DOG

Rose—Hips.—The ripe fruit of the Dog Rose, Rosa canina, and other indigenous allied species.

Characters.—An inch or more in length, ovate, scarlet, smooth, shining ; taste sweet, subacid, pleasant.

Preparation.

Confectio Rosæ Caninæ.—1 in 3.   *Dose,* 60 gr. or more.

Confectio Rosæ Caninæ is contained in Pilula Quiniæ.

ACTION AND USE.

The confection of hips forms a very useful **basis for** pills.

Composition.—Hips contain *malic* and *citric acids,* free and combined, *tannic acid, sugar,* and a trace of *volatile oil.*

Amygdala Dulcis—Sweet Almond.—The seed of the sweet almond tree, Amygdalus communis, *var.* dulcis.   Cultivated about Malaga.

Characters.—Above an inch in length, lanceolate, acute, with a clear cinnamon-brown seed-coat, and a bland sweetish nutty-flavoured kernel.   Does not evolve the odour of bitter almonds when bruised with water.

Impurity.—The bitter almond, which yields an odour of hydrocyanic acid when bruised with water.

Amygdala Amara— Bitter Almond. — The seed of the bitter almond tree, Amygdalus communis, *var.* amara.   Brought chiefly from Mogadore.

Characters.—Resembles the sweet almond in appearance, but is rather broader and shorter ; has a bitter taste, and when rubbed with a little water emits a characteristic odour.

Composition.—Both varieties of almond yield by expression about 50 per cent. of *fixed* oil, Oleum Amygdalæ, and albuminous substances, including *emulsin.*   The bitter variety also yields, by distillation with water, a *volatile* oil, Oleum Amygdalæ Amara, Essential Oil of Almonds, not officinal.

This Essential Oil of Bitter Almonds, or "Oil of Bitter Almonds," although not officinal, must be carefully distinguished from the official fixed oil, Oleum Amygdalæ, inasmuch as in the crude form generally sold it is highly poisonous, from admixture with 4 to 8 per cent. of hydrocyanic acid. Bitter almonds contain neither the volatile oil nor hydrocyanic acid until moistened, but 2 to 3 per cent. of a body called *amygdalin*, $C_{20}H_{27}NO_{11}$, a crystalline glucoside, which, in the presence of water, and under the fermentive influence of the emulsin, breaks up into the volatile oil, hydrocyanic acid, and glucose : $C_{20}H_{27}NO_{11} + 2H_2O = C_7H_6O + HCN + 2C_6H_{12}O_6$. When purified by separation of the hydrocyanic acid, volatile oil of bitter almonds is not poisonous, consisting, as it does, of hydride of benzol (C_7H_5OH), with benzoic acid ($C_7H_6O_2$) as a product of oxydation by exposure, and other allied substances, and is used for flavouring sweets. Nitro-benzine, however, which is sometimes substituted for it, having a very similar flavour, is decidedly poisonous.

Preparations of the Sweet Almond.

1. **Oleum Amygdalæ.**—Almond Oil. The oil *expressed* from bitter and sweet almonds. Pale yellow, nearly inodorous or with a nutty odour, and a bland oleaginous taste. *Dose*, 2 to 4 fl.dr.

   Almond Oil is contained in Unguentum Cetacei, Unguentum Simplex (and its preparations), Unguentum Hydrargyri Oxidi Rubri, Unguentum Plumbi Subacetatis Compositum. It is used in preference to olive oil, as it makes a whiter ointment.

2. **Pulvis Amygdalæ Compositus.**—8 to 4 of Sugar, and 1 of Gum Acacia. *Dose*, 60 to 120 gr.

   Pulvis Amygdalæ is used in preparing :
   a. **Mistura Amygdalæ.**—1, with water, 8. *Dose*, 1 to 2 fl.oz.

ACTION AND USES.

The Sweet Almond is **demulcent and nutritive**, and has been ground into a flour for making cakes to be eaten by diabetic patients, instead of starchy food. The Compound Powder and Mixture are used only as vehicles for insoluble powders and demulcent cough medicines.

Almond oil has the same action, and is used for the same purposes, as olive oil, which, though less agreeable, is more generally employed as being cheaper. See *Oleum Olivæ*, p. 284.

Prunum—PRUNE.—The dried drupe of the Plum, Prunus domestica. From southern Europe.

Composition.—The prune contains *sugar, malic acid,* and a purgative principle.

Prune is contained in Confectio Sennæ, 1 in 12½.

ACTION AND USES.

The prune is **nutritive, demulcent, and slightly laxative,** It may be ordered as an article of diet in habitual constipation.

Laurocerasi Folia—CHERRY-LAUREL LEAVES. —The fresh leaves of Prunus Laurocerasus. The Common or Cherry Laurel.

Characters.—Ovate-lanceolate or elliptical, distantly toothed, furnished with glands at the base, smooth and shining, deep green, on strong short footstalks; emitting a ratafia odour when bruised.

Composition.—Cherry laurel leaves yield by distillation a variable amount of *hydrocyanic acid,* and a *volatile oil,* by a process of decomposition resembling that just described in the bitter almond. Neither emulsin nor ordinary amygdalin have, however, been demonstrated in the leaves, but a resinoid body, which yields with emulsin hydrocyanic acid, and is called "*amorphous amygdalin.*"

Preparation.

Aqua Laurocerasi,—1 in 1¼ by distillation. *Incompatibles,* metallic salts. *Dose,* 5 to 30 min.

ACTION AND USES.

Cherry-laurel water possesses the action of hydrocyani acid, and is also a **flavouring** agent. The strength of the drug in this very powerful substance is so uncertain that **its use** **ought to be avoided.** See *Acidum Hydrocyanicum Dilutum.*

Cusso—Kousso.—The flowers and tops of Brayera anthelmintica. Collected in Abyssinia.

Characters.—Flowers small, reddish brown, on hairy stalks, outer limb of calyx five-parted, the segments oblong or oblong-lanceolate, reticulated, with a fragrant odour and a disagreeable acrid taste.

Composition.—Kousso contains a *volatile oil,* tannic acid, gum, sugar, and a neutral crystallisable active principle, *koussin*, or cossin, $C_{31}H_{38}O_{10}$.

Dose.—$\frac{1}{4}$ to $\frac{1}{2}$ oz.

Preparation.

Infusum Cusso.—$\frac{1}{4}$ oz. in 4 fl.oz. boiling water for one dose; to be drunk without straining.

ACTION AND USES.

Taken in the large doses necessary, kousso is apt to cause nausea, vomiting, colic, and slight diarrhoea. Its principal action is as an **anthelmintic**, the tape-worms (Tænia solium, Tænia mediocanellata, and Bothryocephalus latus) being readily killed by it. It is used for this purpose only, and rarely in England. It may or may not require the assistance of a purgative to expel the dead worm. The powdered flowers, either in compressed masses or suspended in an aromatic water, are said to be much more active than the officinal infusion.

MYRTACEÆ.

Caryophyllum—Cloves. — The dried unexpanded flower buds of Caryophyllus aromaticus. Cultivated in Penang, Bencoolen, and Amboyna.

Characters.—About six lines long, dark reddish-brown, plump and heavy, consisting of a nearly cylindrical body surmounted by four teeth and a globular head, with a strong fragrant odour, and a bitter spicy pungent taste. It emits oil when indented with the nail.

Composition.—Cloves contain 20 per cent. of the officinal *oil*, tannic acid, and gum. Oil of cloves consists of *eugenol* (eugenio acid), $C_{10}H_{12}O_2$, chemically resembling phenol (carbolic acid), and a turpentine. A crystalline body, *eugenin*, isomeric with eugenol, and a neutral body, *caryophyllin*, isomeric with camphor, can also be obtained from cloves.

Preparations.

1. **Oleum Caryophylli.**—Oil of cloves. The oil distilled in Britain from cloves. Colourless, when recent, becoming red-brown, with the odour and burning spicy taste of the clove. It is one of the few volatile oils heavier than water. *Dose,* 1 to 4 min.

2. **Infusum Caryophylli.**—1 in 40. *Dose,* 1 to 2 fl.oz.

   Cloves and Oil of Cloves are also contained in several preparations of other drugs.

ACTION AND USES.

Cloves may be taken as the type of a great group of remedies, other members of which are orange, lemon, pimenta, cajuput, carui, dill, peppermint, and many more, which are met with in our systematic review of medicinal plants. This group is known as the **aromatic essential oils,** of complex and variable chemical composition, but consisting as a rule of terpenes, mixed with camphors, resins, fatty and other acids, and different vegetable constituents. They are closely allied, on the one hand, to phenol (carbolic acid) and benzoic acid, and on the other to still more complex vegetable products, the balsams and gum-resins. Instead of dislocating the various members of the group of aromatic oils from their proper botanical position to discuss them together, we will describe their action and uses once for all under the present head, it being understood that what is said of oil of cloves applies to the other substances, with insignificant qualifications.

1. IMMEDIATE LOCAL ACTION AND USES.

Externally, the essential oil of cloves and allied substances closely resemble turpentine in properties. Thus, whilst **preventing or arresting** decomposition, they redden and inflame the skin, and cause for a time smarting pain, which gives place to local anæsthesia. Oil of cloves and other officinal fragrant oils are too costly to be used externally, except to scent liniments; but the concrete " oils," or solid constituents of the oils, of peppermint, thyme, eucalyptus, myrtle, etc. (stearoptenes), are excellent **antiseptics, local anæsthetics, stimulants** and **counter-irritants** and turpentine and camphor are common applications for these purposes. Such aromatic substances might be used to disinfect foul wounds and ulcers, and promote healing; to hasten the removal of chronic inflammatory products by increasing the local blood-flow, and thus to reduce swelling in or under the skin, the periosteum, or the joints; to relieve neuralgic and rheumatic pains, such as sciatica and lumbago, by dulling the sensibility of the nerves; and to act reflexly on deeper parts, for instance, the lung or heart, when applied to the skin over them as counter-irritants.

Internally.—In the mouth the aromatic oils of cloves and its allies act much as they do on the skin. Besides being antiseptic, they dilate the local vessels (? directly), and thus increase the circulation, heat, and nutrition, and may even cause inflammation. They irritate the nerves, causing pain associated with a sense of burning; but depression quickly follows, and local anæsthesia. Oil of cloves is a valuable application in toothache from dental caries, acting at once as an anodyne and disinfectant. At the same time, the nerves of **taste** and **smell** (flavour) are powerfully excited. Several results, of the first importance in digestion, follow these local changes, namely : (1) reflex salivation; (2) reflex flow of mucus; (3) reflex hyperæmia of the gastric mucosa, a sense of hunger, and a flow of gastric juice; (4) stimulation of the appetite and increase of relish by the pleasing flavour; and (5) in a word, **increased desire for, enjoyment of, and digestion of food.**

Aromatic oils are accordingly used very extensively in cookery, where the proper use of them constitutes an important portion of the culinary art. Those of them which are also bitter, such as orange, are taken with wines and spirits as various "aromatic bitters," liqueurs, etc., to rouse or strengthen appetite and digestion before or during a meal. In pharmacy they are employed to correct the tastes of nauseous drugs; and therapeutically they are given in dyspepsia and debility along with most bitters to increase the saliva and gastric juice.

In the stomach, the effect of aromatics on the vessels and nerves is continued; and here it is generally described as **carminative.** Besides causing an increased flow of juice, by stimulation of the mouth, these substances are powerful **stomachics** in several ways. The vessels of the mucosa are dilated; the nerves of the same are first excited (causing a sense of heat in the epigastrium) and then soothed, thus relieving pain ; the contents, if decomposing, as in dyspepsia, are partly disinfected. Their reflex influence is equally important. The muscular coat is stimulated, thus increasing the gastric movement, expelling flatulence, and relieving painful cramps, spasms, hiccup, and other forms of distress. Distant organs are also stimulated : the vigour of the heart increased, the blood pressure raised, and the spinal, medullary, and even cerebral centres temporarily excited, to the relief of low, hysterical, and "spasmodic" symptoms, very common in certain classes of females, as well as of more serious conditions, such as asthma, cardiac pain, and palpitation. Aromatics are thus **general stimulants** and **antispasmodics.**

In the intestines the aromatic oils may still be found partly unabsorbed, acting on the same structures as before, increasing the local functions, stimulating the intestinal movements, and expelling flatus. They thus relieve or prevent pain or spasm (colic), and provide us with valuable **correctives** of the griping tendencies of many purgatives. The constitution of the most important compound pills, powders, and laxative draughts should be studied in this connection, such as Pilula Rhei Composita, Pulvis Jalapæ Compositus, and Mistura Sennæ Compositus. Caryophyllum is slightly astringent, by virtue of its tannic acid.

2. ACTION ON THE BLOOD.

The aromatic oils of cloves and its allies enter the blood as such, and whilst oxydised in part by the red corpuscles, leave the circulation mainly unchanged. Some of them are known to increase the **number of white corpuscles,** possibly by acting on the lymphatic glands or spleen.

3. SPECIFIC ACTION AND USES.

The aromatic oils are rarely given in sufficient doses to produce definite specific effects on the tissues and organs. It may safely be assumed that in the main their action closely resembles that of turpentine, or that of camphor, respectively, as the one or the other compound is in excess in the particular drug. (*See* these substances.) Speaking generally, they are stimulant and antispasmodic; but let it be noted that a great part of this effect is reflex from the stomach, as described.

4. REMOTE LOCAL ACTION AND USES.

The aromatic oils are excreted by the kidneys, skin, bronchi, liver, and probably the bowels; and in passing through these structures **stimulate** and **disinfect** them. This subject is of the first importance in pharmacology, and will be best discussed under the head of turpentine, an oil which produces very marked remote effects. See *Terebinthinæ Oleum,* page 346.

Pimenta—PIMENTO.—The dried unripe berries of the Allspice tree, Eugenia Pimenta. West Indies.

Characters.—Of the size of a small pea, brown, rough, crowned with the teeth of the calyx, yellowish within, and containing two dark brown seeds. Odour and taste aromatic, hot, and peculiar.

Substances resembling Pimenta : Pepper, which has no calyx.

Composition.—Pimento contains chiefly the officinal *volatile oil,* identical with oil of cloves.

Preparations.

1. **Oleum Pimentæ.**—The oil distilled from the fruit in England. Colourless, becoming brown by keeping. Sinks in water. *Dose,* 1 to 3 min.
2. **Aqua Pimentæ.**—1 in 11½, by distillation. *Dose,* 1 to 2 oz.

Pimenta is also contained in Syrupus Rhamni.

ACTION AND USES.

The action and uses of pimento are the same as those of the preparations of cloves, and other **aromatics**.

Oleum Cajuputi—OIL OF CAJUPUT.—The Oil distilled from the leaves of Melaleuca minor. Imported from Batavia and Singapore.

Characters.—Very mobile, transparent, of a fine pale bluish-green colour. It has a strong agreeable odour, and a warm aromatic taste, and leaves a sensation of coldness in the mouth.

Composition.—Oil of Cajuput consists of *hydrate of cajuputene* ($\frac{2}{3}$), isomeric with Borneo camphor, $C_{10}H_{18}O$, and a second oil ($\frac{1}{3}$), boiling at a higher temperature.

Impurities.—Copper; detected by usual tests. Other volatile

Dose.—1 to 3 min.

Preparation.

Spiritus Cajuputi.—1 in 50.   *Dose*, 30 to 60 min.

   Oil of Cajuput is also contained in Linimentum Crotonis. $3\frac{1}{2}$ in 8.

ACTION AND USES.

Cajuput oil resembles in its action and uses oil of cloves, just described, but it is more used externally as a **stimulant** and **counter-irritant**.

Eucalypti Folia—EUCALYPTUS LEAVES.   (*Not Officinal.*)—The dried leaves of Eucalyptus globulus, the Blue Gum tree of Australia.

Characters.—Grey-green coriaceous leaves, 6 to 12 inches in length, $\frac{1}{2}$ to 1 inch in breadth, ensiform, smooth, entire, studded with oil glands; smell, camphoraceous; taste, bitter and pungently aromatic.

Composition.—The leaves contain 2·75 to 6 per cent. of an æthereal oil, *eucalyptol*, *resin*, *tannin*, salts, and colouring matter. Eucalyptol is a colourless mobile liquid obtained by distillation from the leaves, with a strong agreeable odour, consisting of 70 per cent. of a terpene, with cymol. It readily changes into resin, yielding ozone.

Dose.—Of the leaves, 5 gr. and upwards; of the oil, 1 to 5 min.

Preparations.

Eucalyptus Gauze; and Eucalyptus Ointment.—1 of the oil in 5.

ACTION AND USES.

Externally.—Eucalyptus oil is a powerful **antiseptic and disinfectant.** The gauze has almost supplanted carbolic acid gauze in Lister's process, as it is neither irritant locally, nor poisonous when absorbed.

Internally.—The action of eucalyptus oil is nearly the same as that of oil of turpentine, with which it is otherwise so closely allied. (See *Terebinthinæ Oleum.*) It is **antipyretic and antiperiodic** to a degree, like quinia, and was once believed to be of great value in ague, but this is now doubtful. The blue gum tree is planted in aguish districts to free the soil of malaria.

The remote local action of eucalyptus is important. It leaves the system by the kidneys and lungs, giving its odour to their excretions, and disinfecting these and the mucous surfaces. The oil is therefore indicated in pyelitis and cystitis on the one hand; and in bronchitis, dilated bronchi, and asthma on the other hand.

Granati Radicis Cortex — POMEGRANATE ROOT BARK.—The dried bark of the root of Punica Granatum. Obtained from the south of Europe.

Characters.—In quills or fragments of a greyish-yellow colour externally, yellow internally, having a short fracture, little odour, and an astringent slightly bitter taste.

Incompatibles.—Alkalies, lime-water, metallic salts, gelatine.

Composition.—Pomegranate root bark contains *tannin;* a crystallisable body *punicin;* a substance resembling mannite (see *Manna*), mucilage, etc.

Preparation.

Decoctum Granati Radicis.—1 in 10. *Dose,* 1 to 2 fl.oz.

ACTION AND USES.

Pomegranate root bark has an **anthelmintic** and slightly irritant action, much like kousso (see *Cusso,* page 242), but is somewhat **astringent** unless taken in large quantities. It has long been used in the treatment of the tape-worm, which is expelled apparently (not actually) dead by a dose of the decoction, preceded and followed by a purgative.

DIPTEROCARPINEÆ.

Chaulmoogra Oil.—The oil expressed from the seeds of Gynocardia odorata. From India.

Characters.—A pale-brownish unctuous solid, with a disagreeable smell and taste.

Composition.—Chaulmoogra oil contains a quantity of *palmitic acid*, with three other fatty acids, including *gynocardic acid*, the supposed active principle.

Dose.—2 to 15 gr.

ACTION AND USES.

Chaulmoogra oil is believed to be **a local stimulant**, and a **nutritive** when administered either by inunction or internally. It was for a time much praised in leprosy, and has been used for phthisis, lupus, psoriasis, and chronic rheumatism.

CUCURBITACEÆ.

Colocynthidis Pulpa—Colocynth Pulp.—The dried decorticated fruit, freed from seeds, of Citrullus Colocynthis. Imported chiefly from Smyrna, Trieste, France, and Spain.

Characters.—Light, spongy, white or yellowish-white in colour, intensely bitter in taste.

Composition.—The active principle of colocynth is a bitter glucoside *colocynthin*, $C_{56}H_{84}O_{23}$, usually amorphous, but crystallisable, and readily soluble in water.

Dose, in powder, 2 to 8 gr.

Preparations.

1. **Extractum Colocynthidis Compositum.**—Colocynth Pulp, 6; Extract of Socotrine Aloes, 12 ; Resin of Scammony, 4 ; Hard Soap, 3 ; Cardamom Seeds, 1 ; Proof Spirit, 160. *Dose*, 2 to 5 gr.
2. **Pilula Colocynthidis Composita.**—Colocynth Pulp, 1 ; Barbadoes Aloes, 2 ; Scammony, 2 ; Sulphate of Potash, ¼ ; Oil of Cloves, ¼ ; Water, q.s. (about ¼). *Dose*, 5 to 10 gr.

3. **Pilula Colocynthidis et Hyoscyami.**—Made like the compound pill with ⅓ its weight of Extract of Hyoscyamus. *Dose*, 5 to 10 gr.

ACTION AND USES.

1. IMMEDIATE LOCAL ACTION AND USES.

Colocynth is a powerful **gastro-intestinal stimulant** or irritant, according to the amount, causing speedy large and watery evacuations of the bowels, attended by griping and general depression unless its effect be covered by a carminative. It is one of the most powerful of officinal purgatives, acting as a **hydragogue cathartic** at once upon the muscular coat and intestinal glands and liver, the secretions of which are rendered abundant and watery.

Colocynth is always used in combination with milder purgatives and carminatives. The compound pills are extensively employed alone, or with calomel or blue pill, as an occasional purgative, to produce free evacuation of the bowels, and relieve the portal system, after free living, bilious derangement, or chronic constipation. It is less suitable as a habitual purgative. Its hydragogue effect is employed in cerebral congestion, where rapid "derivation" is required, and in dropsies, especially ascites, either alone or as the basis of a pill containing elaterium. Colocynth must be given with caution in pregnancy, and entirely avoided in delicate or irritable conditions of the stomach and bowels.   ·

2. ACTION IN THE BLOOD; SPECIFIC AND REMOTE LOCAL ACTION.

Colocynthin may be taken up by the skin, enters the blood, and is excreted partly by the kidneys, being, according to some, a diuretic.

Ecbalii Fructus—SQUIRTING CUCUMBER FRUIT. —The fruit, very nearly ripe, of the Squirting Cucumber, Ecbalium Officinarum.

Composition.—Elaterium contains an active neutral principle, *elaterin*, $C_{20}H_{28}O_5$, occurring in small colourless silky prisms, odourless, with an intensely bitter acrid taste; insoluble in water, soluble in spirit.

Preparation.

Elaterium.—A sediment from the juice of Ecbalium Officinarum.

Source.—Made by pressing the juice from the incised fruit, straining, filtering, and drying the sediment.

Characters.—In flattened or slightly incurved pieces about 1 line thick ; light, greenish-grey, friable.

Impurities.—Starch, flour, and chalk ; detected by ordinary tests.

Dose.—$\frac{1}{16}$ to $\frac{1}{2}$ gr.

From Elaterium is prepared :

Pulvis Elaterii Compositus,—1 to 9 of Sugar of Milk.
Dose, $\frac{1}{2}$ to 5 gr.

ACTION AND USES.

Elaterium acts much like colocynth, as a gastro-intestinal irritant, but is decidedly more violent, being the most powerful hydragogue purgative which we possess. It produces, even in doses of $\frac{1}{12}$ to $\frac{1}{8}$ gr., numerous very watery motions, with griping and considerable depression.

Elaterium is used almost entirely as a hydragogue purgative in dropsies and uræmia, relieving the venous pressure by free evacuation of fluid into the bowel. More rarely it is given as a rapid " derivative " in cerebral cases ; and still more rarely as an evacuant in obstinate constipation. This drug must be used with caution, doses of $\frac{1}{16}$ grain being given at first, as the strength in elaterin is uncertain. It must not be ordered in catarrhal states of the stomach or bowels.

UMBELLIFERÆ.

Conii Folia— Hemlock Leaves. — The fresh leaves and young branches of Spotted Hemlock, Conium maculatum ; also the leaves separated from the branches and carefully dried ; gathered from wild British plants when the fruit begins to form.

Characters.—Fresh leaves decompound, smooth, arising from a smooth stem with dark purple spots ; dried leaves of a full green colour and characteristic odour. The leaf rubbed with solution of potash gives out strongly the odour of conia.

Dose, in powder.—2 to 8 gr.

Conii Fructus—Hemlock Fruit.—The dried ripe fruit of Conium maculatum, Spotted Hemlock.

Characters.—Broadly ovate, compressed laterally ; half-fruit with five waved or crenated ridges. Reduced to powder and rubbed with solution of potash, they give out strongly the odour of conia.

Substances resembling Conium Fruit : Caraway, Anise, Dill, known by presence of vittæ.

Composition.—The active principle of conium is a liquid alkaloid, *conia,* $C_8H_{15}N$. It is strongly alkaline, oily, and volatile ; and has a peculiarly disagreeable mouse-like odour. It is readily disengaged from the preparations of the plant by the addition of alkalies ; and is liable both to conversion into an inert resinous mass by exposure, and to decomposition by heat. The preparations of conium, for these and probably other reasons, are peculiarly uncertain in strength and action. *Coniic acid*, and a second alkaloid *conhydrin*, also exist in hemlock.

Incompatibles.—Caustic alkalies, vegetable acids, and astringents.

Preparations.

A. *Of Conii Folia :*

1. **Cataplasma Conii.**—1 oz. of the *dried* leaf powdered in each.
2. **Extractum Conii.**—A green extract from *fresh* leaves. About 30 in 1. *Dose,* 2 to 6 gr.

From Extract of Conium are prepared :

   a. Pilula Conii Composita. — Extract of Hemlock, 5 ; Ipecacuanha, 1 ; Treacle, q.s. *Dose,* 5 to 10 gr.
   b. Vapor Conii.—Extract of Hemlock, 60 gr. ; Solution of Potash, 1 fl.dr. ; Water, 10 fl.dr. 20 min. for one inhalation.

3. **Succus Conii.**—3 of the expressed juice of the *fresh* leaves, with 1 of Spirit. *Dose,* 30 to 60 min. (B. Ph.)

B. *Of Conii Fructus :*

Tinctura Conii.—1 in 8. *Dose,* $\frac{1}{2}$ to 1 fl.dr.

ACTION AND USES.

1. IMMEDIATE LOCAL ACTION AND USES.

Externally applied, as the cataplasm, conium is believed by many to be **anæsthetic** and especially to relieve the pain of cancer, as well as to promote the absorption of tumours. Careful experiment fails to confirm this opinion, the whole of the sensory nervous system remaining unaffected by the drug, except indirectly by poisonous doses.

Internally. — Conium sometimes causes irritation and vomiting.

2. ACTION ON THE BLOOD.

Conia is readily absorbed into the blood, whence it reaches the tissues.

3. SPECIFIC ACTION AND USES.

Conia is found unchanged in many of the organs after administration. Moderate doses cause a sense of weight in the legs and weakness of the knees; confusion of vision, with slight drooping of the upper lids, and swollen appearance of the eyes; giddiness, thickness of speech, and slight dysphagia. The poisonous effects of the plant are well described in the classical account of the death of Socrates.

On analysis, the action of conium is found to be as follows. The convolutions remain intact until asphyxia supervenes. The corpora striata are said to be depressed. The motor parts of the cord are but slightly affected, but their reflex excitability is moderately reduced. The **respiratory** centre in the medulla is finally **paralysed**; but the cardiac and vascular centres are not definitely influenced.

The **motor nerves are** the parts specially attacked by conium, being **paralysed from** their **extremities upwards,** whence the heaviness and weakness of the limbs. The muscles themselves remain irritable.

Death occurs in hemlock poisoning by asphyxia due to paralysis of the respiratory nerves and depression of the respiratery centre.

Conium, although of great interest to the pharmacologist, is but little used in medicine. It has been recommended, as large doses of the succus, in spasmodic and convulsive diseases such as tetanus, chorea, and epilepsy; in mania with muscular excitement; and in asthma, pertussis, and spasmodic affections of the larynx. The vapour would appear to afford relief in some of the last-named class of cases. Possibly the compound pill may allay spasmodic cough. The extract is an adjuvant vehicle of purgative powders such as calomel.

4. REMOTE LOCAL ACTION.

Conia is excreted unchanged, chiefly in the urine.

Assafœtida—ASSAFŒTIDA.—A gum-resin obtained by incision from the living root of Narthex Assafœtida. In Afghanistan and the Punjaub.

Characters.—In irregular masses, partly composed of tears, moist or dry. The colour of a freshly cut or broken piece is opaque white, but gradually becomes purplish-pink, and ultimately dull yellowish or pinkish-brown.   Taste bitter, acrid; odour fetid, alliaceous, and persistent.   It dissolves almost entirely in rectified spirit.

Composition.—Assafœtida contains 4 per cent. of a *volatile oil*, 65 per cent. of *resin*, and 25 per cent. of *gum*.   Oil of assafœtida is probably complex, but consists chiefly of sulphide of allyl, $C_6H_{10}S$, to which the unpleasant odour is due.   The resin also contains sulphur.

Impurities.—Earthy matter.

Substances resembling Assafœtida : Galbanum, Ammoniacum, Benzoin ; known by odour.

Dose.—5 to 20 gr.

Preparations.

1. <u>Enema Assafœtidæ.</u>—30 gr. in 4 fl.oz. of water.
2. <u>Pilula Aloes et Assafœtidæ,</u>—1 in 4.   (*See Aloes*, page 359.)
       Dose, 5 to 10 gr.
3. <u>Pilula Assafœtidæ Composita.</u>—Syn.: Pilula Galbani Composita.   Assafœtida, 2 ; Galbanum, 2 ; Myrrh, 2 ; Treacle, by weight, 1.   *Dose*, 5 to 10 gr.
4. <u>Spiritus Ammoniæ Fœtidus.</u>—Assafœtida, 1½ ; Strong Solution of Ammonia, 2 ; Spirit, 20.   *Dose*, ½ to 1 fl.dr.
5. <u>Tinctura Assafœtidæ.</u>—1 in 8.   *Dose*, ½ to 1 fl.dr.

ACTION AND USES.

1. IMMEDIATE LOCAL ACTION AND USES.

Assafœtida possesses the action of other volatile oils and resin upon the alimentary canal, but differs from them in this highly important respect, that whilst most of them are aromatic and pleasant to the palate, it is extremely disagreeable. The mental effect of this nauseous impression, added to the other stimulant effects on the mouth and stomach (see *Caryophyllum*, page 242), constitute assafœtida a powerful nervine stimulant, which arrests the emotional disturbance, muscular spasms, and other morbid nervous disorders of hysteria.   It is no longer used in true epilepsy, chorea, laryngismus, or asthma.   The stimulant action of volatile oils on the bowel (see *Terebinthinæ Oleum*, page 343) is specially marked and is employed in the Enema Assafœtidæ to expel flatulence, relieve constipation, and arrest convulsions.

2. ACTION IN THE BLOOD ; SPECIFIC, AND REMOTE LOCAL ACTION AND USES.

The volatile oil of assafœtida passes through the blood and tissues, and is excreted in the urine, sweat, breath, and discharge from wounds. Thus remotely it exerts the usual stimulant action of ethereal oils, and is sometimes given as a stimulant and disinfectant expectorant in chronic bronchitis.

Galbanum—GALBANUM.—A gum-resin, derived from an unascertained umbelliferous plant (said to be Ferula galbaniflua). Imported from India and the Levant.

Characters.—In irregular tears, about the size of a pea, usually agglutinated into masses; of a greenish-yellow colour, translucent; having a strong disagreeable odour, and an acrid bitter taste.

Substances resembling Galbanum: Ammoniacum, Assafœtida, Benzoin; known by odour.

Composition.—Galbanum contains 3 to 6 per cent. of *volatile oil*, isomeric with turpentine, $C_{10}H_{16}$, *gum*, and a mixture of *resins*, which yield by dry distillation a blue oil, and *umbelliferon*, in colourless, tasteless, odourless, satiny crystals.

Preparations.

Emplastrum Galbani.—1 in 11.

Galbanum is also an ingredient of Pilula Assafœtidæ Composita. 1 in 3½. See *Assafœtida*, page 252.

ACTION AND USES.

Galbanum acts and is used much **like assafœtida** and ammoniacum, and is always given with either of these substances.

Ammoniacum—AMMONIACUM.—A gum-resinous exudation from Dorema Ammoniacum. Collected in Persia and the Punjaub.

Characters.—In tears or masses; the tears from two to eight lines in diameter, pale cinnamon-brown, breaking with a smooth shining opaque white surface; the masses composed of agglutinated tears; hard and brittle when cold, but readily softening with heat. Has a faint odour, and a bitter acrid nauseous taste. Rubbed with water it forms a nearly white emulsion.

Substances resembling Ammoniacum : Assafœtida, Galbanum, Benzoin; known by odour.

Composition.—Ammoniacum contains about 4 per cent. of a

volatile oil, 20 per cent. of *gum*, and 70 per cent. of *resin*. The oil does not contain sulphur.

   Dose.—10 to 20 gr.

Preparations.

1. **Emplastrum Ammoniaci cum Hydrargyro.**—About 1 in 1¼. (See *Hydrargyrum*, page 86.)
2. **Mistura Ammoniaci.**—A milk-like emulsion. 1 in 32 water. *Dose,* ½ to 1 fl.oz.

Ammoniacum is also an ingredient of Emplastrum Galbani, 1 in 11; Pilula Ipecacuanhæ cum Scillâ, 1 in 7; and Piïula Scillæ Composita, 1 in 6.

ACTION AND USES.

The action of ammoniacum closely resembles that of the other aromatics and oleo-resins, but it is used almost solely for its remote local effects. In being excreted by the bronchial mucosa, it stimulates the surface and disinfects the secretions of the part (see *Terebinthinæ Oleum*, page 343); and it probably acts similarly on the skin. It is used as a **disinfectant expectorant** in chronic bronchitis with profuse discharge, and as a constituent of plasters intended to strengthen circulation in the skin and promote absorption.

Oleum Anisi—OIL OF ANISE.—The oil distilled in Europe from the fruit of Pimpinella Anisum. Also the oil distilled in China from the fruit of Illicium anisatum, Star Anise (Magnoliaceæ.)

Characters of the fruit.—Half fruits with five filiform equal ridges, the lateral ones being marginal. In each channel are three or more vittæ. Thicker and more ovate than caraway fruits.

Characters of the oil.—Colourless or pale yellow, with the familiar odour of anise, and a warm sweetish taste. Concretes at 50°.

Composition.—Oil of aniseed, *anethol*, or anise-camphor, is composed of two isomeric bodies, the *fluid* (⅕), and the *solid* (⅘), anethol.

   Dose.—2 to 5 min.

Preparation.

Essentia Anisi.—1 in 5. *Dose,* 10 to 20 min.
Oil of Anise is also contained in Tinctura Camphoræ Composita, and Tinctura Opii Ammoniata.

ACTION AND USES.

The action and uses of anise are those of the **aromatic** oils in general. It is believed, however, to possess specially stimulant action on the bronchial mucosa, like ammoniacum, probably because excreted in part by it. It is therefore a favourite flavouring agent for cough mixtures.

Coriandri Fructus—CORIANDER FRUIT.—The dried ripe fruit of Coriandrum sativum. Cultivated in Britain.

Characters.—Globular, nearly as large as white pepper, beaked, finely ribbed, yellowish-brown; has an agreeable aromatic odour and flavour.

Composition.—The principal constituents of coriander are an abundant *fatty oil*, and a small quantity of *aromatic oils*, one of which is isomeric with Borneo camphor, $C_{10}H_{18}O$. See *Camphora.*

Preparation.

Oleum Coriandri.—Obtained by distillation. Yellowish, with the odour of coriander. *Dose*, 2 to 5 min.

Coriander Fruit and Oil are also contained in a variety of preparations of more important drugs.

ACTION AND USES.

The action and uses of coriander do not differ from those of the **aromatic** substances just described. Its flavour specially covers the taste of senna and rhubarb.

Fœniculi Fructus—FENNEL FRUIT.—The fruit of Fœniculum dulce. Imported from Malta.

Characters.—About three lines long and one line broad; elliptical, slightly curved, beaked, having eight pale-brown longitudinal ribs, the two lateral being double; taste and odour , aromatic.

Substances resembling Fennel: Conium, Caraway, Anise. Fennel is larger than Conium, and has eight ribs and often footstalk.

Composition.—Fennel contains an ethereal *volatile oil*, apparently identical with anethol (see *Oleum Anisi*, page 255;) united with a terpene. It is of a light yellow colour, with the peculiar odour of the fruit.

Preparation.

Aqua Fœniculi.—1 in 10 by distillation. *Dose,* 1 to 2 fl.oz.

ACTION AND USES.

Fennel has the same action, and is used for the same purposes as other **aromatic** substances.

Carui Fructus—CARAWAY FRUIT.—The dried fruit of Carum Carui. Cultivated in England and Germany.

Characters.—Fruit usually separating into two parts, which are about two lines long, curved, tapering at each end, brown, with five paler longitudinal ridges; having an agreeable aromatic odour and spicy taste.

Substances resembling Caraway : Conium, Fennel. Caraway has small ridges and a spicy taste.

Composition.—Volatile oil of caraway, the active constituent of the fruit, is a mixture of *caruen*, $C_{10}H_{16}$, isomeric with turpentine, and *caruol*, $C_{10}H_{11}O$, isomeric with thymol.

Preparations.

1. **Aqua Carui.**—1 in 10. *Dose,* 1 to 2 fl.oz.
2. **Oleum Carui.**—Obtained by distillation. Pale yellow, with aromatic odour and spicy taste. *Dose,* 2 to 5 min.

Caraway Fruit and Oil are also contained in many preparations of other drugs.

ACTION AND USES.

Caraway acts like other **aromatic** substances, as described under *Caryophyllum.* It is extensively used as a flavouring and carminative agent.

Anethi Fructus—DILL FRUIT.—The fruit of Anethum graveolens. Cultivated in England, or imported from middle and southern Europe.

Characters.—Oval, flat, about a line and a half in length, with a pale membranous margin. Odour aromatic; taste warm, somewhat bitter.

R—8

Substances resembling Dill : Conium, Anise, Fennel, Caraway. Dill is winged.

Composition.—Dill contains the officinal *volatile oil*, which is pale yellow, with a pungent odour, and an acrid sweetish taste.

Preparations.

1. **Aqua Anethi.**—1 in 10. *Dose*, 1 to 2 fl.oz.
2. **Oleum Anethi.**—Obtained by distillation. *Dose*, 2 to 5 min.

ACTION AND USES.

The same as of other substances containing **aromatic** volatile oils. It is often given as a carminative to infants.

Sumbul Radix—SUMBUL ROOT.—The dried transverse sections of the root of a plant the botanical history of which is unknown. Imported from Russia, and also from India.

Characters.—The pieces are nearly round, from $2\frac{1}{2}$ inches to 5 inches in diameter, and from $\frac{3}{4}$ to $1\frac{1}{2}$ inches in thickness. They are covered on the outer edge with a dusky brown rough bark, frequently beset-with short bristly fibres. The interior is porous, and consists of irregular, easily separated fibres. It has a strong odour, resembling that of musk. The taste is at first sweetish, becoming after a time bitterish and balsamic. That brought from India differs from the Russian, being closer in texture, more dense and firm, and of a reddish tint.

Composition.—Sumbul contains a small quantity of *volatile oil*, 9 per cent. of a soft *resin* with its characteristic odour, and a crystalline substance, *sumbulic acid.*

Preparation.

Tinctura Sumbul.—1 in 8. *Dose*, 15 to 30 min.

ACTION AND USES.

Sumbul is a stimulant, like the **aromatic** oils in general, and specially resembles valerian and musk. It is used in the same class of cases as these drugs, to the account of which the reader is referred.

CAPRIFOLIACEÆ.

Sambuci Flores—ELDER FLOWERS.—The fresh flowers of Sambucus nigra. From indigenous plants.

Characters.—Flowers small, white, fragrant, in cymes.

Composition.—Elder flowers contain a trace of *volatile oil,* a *resin,* and *baldrianic* (valerianic) *acid,* $C_5H_{10}O_2$.

Preparation.

Aqua Sambuci.—1 in 1, by distillation.   *Dose,* 1 to 2 fl.oz.

ACTION AND USES.

Elder flowers are chiefly used for **flavouring** purposes, but probably possess mild diaphoretic and diuretic properties.

CINCHONACEÆ.

Cinchonæ Flavæ Cortex—YELLOW-CINCHONA BARK.—The bark of Cinchona Calisaya.   Collected in Bolivia and southern Peru.

Characters.—In flat pieces, uncoated or deprived of the periderm, rarely in coated quills, from six to eighteen inches long, one to three inches wide, and two to four lines thick, compact and heavy; outer surface brown, marked by broad shallow irregular longitudinal depressions; inner surface tawny-yellow, fibrous : transverse fracture shortly and finely fibrous.   Powder cinnamon-brown, somewhat aromatic, persistently bitter.

Cinchonæ Pallidæ Cortex—PALE-CINCHONA BARK.—The bark of Cinchona Condaminea, D.C., *vars.* Chahuarguera *Pavon,* and crispa *Tafalla.* From Loxa.

Characters.—From half a line to a line thick, in single or double quills, from 6 to 15 inches long, 2 to 8 lines in diameter, brittle, easily splitting longitudinally, and breaking with a short transverse fracture ; outer surface brown and wrinkled, or grey and speckled with adherent lichens, with or without numerous transverse cracks ; inner surface bright orange or cinnamon-brown ; powder pale brown, slightly bitter, very astringent.

Cinchonæ Rubræ Cortex—RED-CINCHONA BARK. —The bark of Cinchona succirubra.   Collected on the western slopes of Chimborazo.

Characters.—In flat or incurved pieces, less frequently in quills, coated with the periderm, varying in length from a few inches to two feet, from one to three inches wide, and two to six lines thick, compact and heavy; outer surface brown or reddish-brown, rarely white from adherent lichens, rugged or wrinkled longitudinally, frequently warty, and crossed by deep transverse cracks; inner surface redder; fractured surface often approaching to brick-red; transverse fracture finely fibrous; powder red-brown; taste bitter and astringent.

Cinchona Lancifoliæ Cortex—LANCE-LEAVED CINCHONA BARK.—The bark of Cinchona lancifolia, *Mutis.* Spongy or orange Carthagena bark.

Characters.—Either in quills of various size with brownish epidermis, and whitish crustaceous and foliaceous lichens, extremely fibrous, moderately bitter; or as curved pieces, of an orange or red colour, with an extremely fibrous liber, of stringy fracture, very slightly bitter.

Composition.—Cinchona bark contains (1) four *alkaloids*, namely: *quinia, cinchonia, quinidia,* and *cinchonidia;* (2) two *peculiar acids : kinic* and *kinovic* acids *;* (3) a variety of tannic acid, called *cincho-tannic acid ;* (4) *cinchona red ;* and (5) an aromatic *volatile oil.*

1. **The alkaloids of cinchona.**—*a. Quinia,* $C_{20}H_{24}N_2O_2$, occurs (as the hydrate) in white acicular crystals, inodorous, very bitter; reacting like an alkali, and forming neutral and acid salts with acids; presenting fluorescence in dilute solutions of the sulphate; and turning the plane of polarisation to the left. An *amorphous* form of quinia is found after crystallisation of the sulphate from the mother-liquors, and from *quinoidia,* which appears to be a compound of the alkaloids with resin and colouring matters.

b. Cinchonia, $C_{20}H_{24}N_2O$, consists of colourless prisms, inodorous, and bitter; forms salts with acids; but possesses no fluorescence in solution; and deflects the plane of polarisation to the right.

c. Quinidia, $C_{20}H_{24}N_2O_2$, *i.e.* isomeric with quinia, closely resembles quinia, but crystallises in prisms, and deflects the plane of polarisation to the right.

d. Cinchonidia, $C_{20}H_{24}N_2O$, *i.e.* isomeric with cinchonia, resembles that alkaloid, but yields fluorescent solutions, and left-handed polarisation.

As a rule quinia is most abundant in yellow bark, cinchonia in pale bark, and the red bark contains a considerable proportion of each. Quinidia is specially abundant in the bark of lancifolia. More exactly, yellow bark yields 2·5 to 3·8 per cent. of quinia; pale bark, 0·7 to 1·4 per cent. of alkaloids, chiefly cinchonia or quinidia with a little quinia; the best red bark, 2·6 per cent. of quinia, and 1·5 per cent. of cinchonia.

2 and 3. **The acids of cinchona.**—*a. Kinic* or *quinic* acid, $C_7H_{12}O_6$, occurs in large colourless prisms, soluble in water. In the bark it is probably combined with the alkaloids, and is found also in the coffee-bean, the Vaccinium myrtillus, and other plants. It is closely allied to benzoic acid, and appears in the urine as hippuric acid. See *Benzoinum*, page 281.

b. Kinovic acid, $C_{24}H_{38}O_4$, "kinova bitter," is a white amorphous body, insoluble in water. It appears to be a product, with glucose, of *kinovin*, a glucoside.

c. Cincho-tannic acid, the astringent principle and soluble red-colouring matter of the bark, amounts to 1 to 3 per cent. It is a yellow hygroscopic body, and differs from ordinary tannic acid in striking green with persalts of iron, and in being very readily oxydised, one of the products being

4. **Cinchona red,** a reddish-brown substance without taste or smell, nearly insoluble in water.

5. **The volatile oil,** obtained by distillation, has the odour of the bark.

Impurities.—Inferior barks are detected by the absence of the true characters of the officinal barks, and by a quantitative test. This consists in (1) boiling 100 gr. of the bark in water acidulated with HCl, macerating, and percolating; (2) precipitating the colouring matter with solution of subacetate of lead; (3) adding caustic potash to the filtrate, until the precipitate first formed is nearly redissolved; (4) agitating with ether, and evaporating the resulting solution. This should yield not less than 2 gr. of quinia from yellow bark; 2 gr. of alkaloids from red bark; and ½ gr. of alkaloids from pale bark. The yellow bark is adulterated with elm, larch, and Winter's barks, known by absence of bitter taste; the pale bark with cascarilla, which is whiter; the red bark with red sandal wood, logwood, and larch barks, which are all devoid of a bitter taste.

Incompatibles.—Ammonia, lime-water, metallic salts, and gelatine. May be combined with mineral acids.

Dose of any of the barks.—15 gr. as a tonic; 1 to 2 dr. in ague.

Preparations.

A. *Of the Yellow Bark :*

1. **Decoctum Cinchonæ Flavæ.**—1 in 16.   *Dose,* 1 to 2 fl.oz.
2. **Extractum Cinchonæ Flavæ Liquidum.**—4 in 1.   *Dose,* 10 to 30 min.

3. **Infusum Cinchonæ Flavæ.**—1 in 20.   *Dose*, 1 to 2 fl.oz.
4. **Tinctura Cinchonæ Flavæ.**—1 in 5.   *Dose*, ½ to 2 fl.dr.
5. **Quiniæ Sulphas.**

    B. *Of the Pale Bark* (*Cinchona Pallida*) :

1. **Tinctura Cinchonæ Composita.**—Pale Bark, Bitter-Orange Peel, Serpentary, Saffron, Cochineal, and Proof Spirit. 1 in 10.   *Dose*, ½ to 2 fl.dr.

    C. *Of the Red Bark* there are no officinal preparations.

    D. *Of the Lance-leaved Cinchona Bark :* Quiniæ Sulphas.

ACTION AND USES.

The action and uses of the Cinchona barks will be described along with those of Quinia, their most important active principle.

Quiniæ Sulphas—SULPHATE OF QUINIA. $(C_{20}H_{24}N_2O_2)_2H_2SO_4.7H_2O$.—The sulphate of an alkaloid, prepared from yellow Cinchona bark, and from the bark of Cinchona lancifolia, *Mutis.*

Source.—Made by (1) exhausting the bark with Diluted Hydrochloric Acid; (2) precipitating the Quinia with Solution of Soda ; (3) dissolving this in Diluted Sulphuric Acid; and (4) crystallising out.

Characters.—Filiform silky snow-white crystals, with an intensely bitter taste. Solubility, 1 in 576 of water : 60 gr. require 60 min. of diluted sulphuric, or 100 min. of diluted phosphoric acid ; and 66 gr. require 60 min. of diluted nitric acid for solution in two ounces of distilled water. 12 gr. = 1 oz. of good bark.

Impurities.—Sulphates of the other cinchona alkaloids, and of lime, chalk, magnesia ; starch, boracic acid, etc.; detected by the quantitative test given above. Salicin; which gives blood-red with sulphuric acid.

Incompatibles.—Alkalies and their carbonates, astringent infusions. Nitric acid alone makes a clear mixture with sulphate of quinia.

Dose.—1 to 5 gr. as a tonic ; larger doses as an antipyretic and antiperiodic.

Preparations.

1. **Ferri et Quiniæ Citras.**—16 in 100.   *Dose*, 5 to 10 gr.
2. **Pilula Quiniæ.**—1 with ⅓ of Confection of Hips.   *Dose*, 2 to 10 gr.

3. **Tinctura Quiniæ.**—1 in 60. *Dose*, 1 to 1½ fl.dr.
4. **Tinctura Quiniæ Ammoniata.**—1 in 60 of Solution of Ammonia and Proof Spirit. *Dose*, ½ to 2 fl.dr.
5. **Vinum Quiniæ.**—1 in 480 of Orange Wine, with Citric Acid. *Dose*, ½ to 1 fl.oz.

ACTION AND USES.

1. IMMEDIATE LOCAL ACTION AND USES.

Externally.—Quinia arrests some kinds of fermentation and decomposition, and might be used as a local **antiseptic and disinfectant** to wounds and ulcers, but for its cost. A solution of 2 gr. to 1 fl.oz., applied as a spray to the nose, relieves hay asthma. A solution of 4 gr. to 1 fl.oz. (with a minimum of diluted sulphuric acid) is recommended as a constant application in diptheritic conjunctivitis, or to wash out a foul bladder.

Internally.—Quinia is freely absorbed by the mucous membranes, and may be given either by the mouth, rectum, or subcutaneously. In the mouth, stomach, and intestine, it acts as a powerful **bitter**, possessing all the important influence on the secretions of the digestive tract described under *Calumba.* The **stomachic** effect of quinia is obtained from small doses, ½ to 2 grains, and must be kept entirely distinct from the specific effects to be presently described, otherwise confusion as to the action and value of this important drug will be the result. In small doses, like all other bitters, it improves the appetite and digestion, stimulates the heart and circulation, increases the sense of comfort and *bien être* produced by a meal; and its continued use will thus increase the bodily strength, that is, will be **tonic** in its effects. Quinia is extensively used for this purpose, especially during convalescence, in debilitated subjects, and in patients taking depressing or alterative remedies such as mercury. Larger doses (10 to 30 gr. or more), have the opposite effect, interfering with digestion, and so causing depression.

In the stomach quinia or its sulphate becomes the chloride, a soluble and diffusible salt, which readily enters the blood. Little or none escapes unabsorbed in the fæces.

2. ACTION ON THE BLOOD, AND ITS USES.

Quinia or its chloride may be found in the blood within a few minutes of its administration. Here the alkaloid produces several definite effects, namely : (1) It binds the oxygen more firmly to the hæmoglobin, so that oxygenation is less easy and less active. (2) It causes enlargement of the individual red corpuscles. (3) It paralyses the leucocytes, when given in large doses, thus checking diapedesis; and reduces the number of visible leucocytes very greatly (to one-fourth). In blood freshly drawn, it (4) retards the formation of acid (through loss of oxygen and increase of carbonic acid) which naturally occurs in blood removed from the vessels, as well as (5) the ozonising power of blood, *e.g.* on guaiacum and turpentine. Altogether, quinia manifestly **interferes with oxygenation**, the giving up of oxygen by the red corpuscles to oxydisable bodies, and with the function of the white corpuscle. The outcome of these effects will be presently considered.

3. SPECIFIC ACTION.

Quinia passes through the tissues without decomposition, quickly making its appearance in them, but not being completely excreted for several days. The maximum effect of large doses is produced in about five hours. If therefore the full specific effect be desired, a single large dose (15 to 30 gr.) must be given, and this may have to be repeated once or twice within the hour; small doses given over a length of time do not sufficiently accumulate.

The obvious phenomena produced by a full dose (15 to 30 gr.) of quinia are not by any means its most important effect. It acts most strikingly upon the nervous centres, and causes confusion of the mental faculties, noises in the ears and deafness, disorders of vision, headache, giddiness, vomiting, and possibly prostration from involvement of the cord and circulation. Of infinitely greater interest and importance are certain concomitant effects of quinia which require careful investigation for their discovery. These effects may be arranged thus:

(1) Quinia **lowers the body temperature** very moderately in the healthy subject; very markedly in the pyrexia of many acute specific fevers. It appears to be difficult to lower the normal temperature by drugs, as compensating mechanisms are probably brought into play; but the rise of temperature and the perspiration normally produced by muscular exercise are prevented by quinia. In malarial fevers, typhoid, acute pneumonia, and some forms of hectic and other periodic fevers, the defervescent effect of quinia is unquestionable.

(2) Quinia **reduces the amount of nitrogenous excretions**, *i.e.* urea and uric acid, and probably also of carbonic acid, as determined both in healthy and in fevered animals, and in man.

These two sets of effects taken together point to a powerful action of quinia in reducing the metabolism of the body, of which heat and the excretions are the two most measurable products. This conclusion is supported by other facts, observed out of the body, viz.: That (3) a solution of albumen cannot be converted into peptones in an atmosphere of ozone if quinia be present. (4) Healthy pus and fresh vegetable juices lose their ozonising power if mixed with quinia. (5) Phosphorescent infusoria (rapidly oxydating protoplasmic masses) lose their phosphorescence in the presence of quinia. (6) Fungi absorb oxygen less readily, and many forms of fermentation are arrested, in the presence of quinia. These facts indicate that quinia so combines with living cellular protoplasm as to render it less able to incorporate oxygen, and more resistant of vital change (metabolism). Now we have already seen that the oxygen actually in the corpuscles is bound more firmly to them by quinia. We may therefore conclude that the effect of quinia in the body is to check metabolism by interfering with oxygenation, with the oxydation of protoplasm generally, and with the associated action of ferments. Thus the fall of temperature produced by quinia is due to diminished production of heat in the body, not to increased loss of heat; it is effected through the tissues, not through the heat-regulating centre; and the fever-causing (pyrogenic) processes themselves (probably allied to fermentations) are also controlled by the drug, which affects their organic causes, whether living organisms or complex chemical substances.

An action such as this upon the processes of nutrition, though it might escape the notice of an ignorant observer, is more "powerful" even than the action of morphia upon a highly sensitive nervous mechanism such as the convolutions.

Turning to the other systems, we find that whilst small doses of quinia accelerate the heart and raise the pressure, as we saw when considering the stomachic effect, full doses slow and weaken the heart and lower the pressure. These effects are due to direct depression of the cardiac ganglia and muscle, and of the vessel walls and their centre, not of the cardiac centre. Respiration is accelerated by medium doses, depressed by large doses; and death, should it occur, is referable to respiratory and cardiac failure. The spleen is reduced in size, and hardened.

4. SPECIFIC USES.

The uses of quinia, which have been mainly established by experience, are in accord with these physiological results. Its specific action may be taken advantage of in the following diseases:

1. *Malaria* is remarkably benefited by quinia, which is an **antiperiodic** or direct specific, whether given to persons exposed to the morbid influence as a prophylactic measure, or to the subjects of ague. It acts best in fresh cases, the first dose of

10 gr. being given at any time with relation to the attack, and similar doses repeated five hours before the time of the next paroxysm. All forms of malarial fever are benefited by quinia, as well as many diseases and disorders of malarial origin, such as neuralgia, hepatic disturbances, etc. The functions of the liver must be maintained during this treatment of ague, and the quinia may be combined with morphia if its effects are not well marked.

2. *Febrile conditions in general* are relieved by the **antipyretic** effect of quinia, for instance, acute pneumonia, typhoid fever, puerperal fever, and septicæmia, the exanthemata, and acute rheumatism; but generally in 'very different degrees, so that its value is denied in some or all of them. To be of use the quinia must be very freely given (10 to 20 grains) as single doses when the temperature reaches a definite height, say 104° Fahr. Even if apyrexia do not follow, the drug may be of much benefit. In hectic fever quinia is rarely of much service; and in purely symptomatic fever, of still less use.

3. In *splenic enlargement* of malarial origin quinia is given with success, and in some cases of leukæmatous hypertrophy.

4. In *painful nervous affections*, especially headache and face-ache, the effect of quinia is well marked. Some of these cases are malarial (brow ague); but ordinary facial neuralgia and toothache will frequently yield to it. Yet quinia possesses no direct action on peripheral nerves.

5. The **tonic** action of quinia has been already referred to. This is also due in part to its removal of fever, and thus of restlessness, sleeplessness, and want of appetite. It further modifies the processes of "secondary digestion" in the liver, and may relieve hepatic disorder due to free living, especially in persons who have resided in the tropics.

5. REMOTE LOCAL ACTION AND USES.

Quinia is excreted chiefly in the urine, as the amorphous alkaloid; partly as resinoid and crystalline derivates. In passing through the urinary organs it is slightly diuretic, and occasionally irritates the passages. It also escapes by the skin, diminishing the perspiration, and very rarely causing an itching eruption, which resembles scarlatina or measles. All the mucous secretions, the milk, and pathological fluids may also contain quinia.

ACTION AND USES OF THE CINCHONA BARKS.

The cinchona barks contain but a small percentage of alkaloids, and are far too bulky for use as antiperiodics and antipyretics if quinia can be obtained. They are therefore given only as **bitter stomachics and tonics.** The amount of tannin contained in them indicates that they may be used when an astringent effect is also desired, either locally, as in passive diarrhœa, or remotely, as in sweating, chronic mucous discharges, and purulent formations; and avoided in constipation, dyspepsia, or irritability of the bowels. The red bark is especially astringent.

Cinchonia, and other non-officinal alkaloids and products of bark, may be employed as substitutes for quinia, their action being very similar. Cinchonia is from $\frac{1}{3}$ to $\frac{1}{2}$ as powerful as quinia.

Ipecacuanha—IPECACUANHA.—The dried root of Cephaëlis Ipecacuanha. Imported from Brazil.

Characters.—In pieces three or four inches long, about the size of a small quill, contorted and irregularly annulated. Colour, brown of various shades. It consists of two parts, the cortical or active portion, which is brittle, and a slender tough white woody centre. Powder pale brown, with a faint nauseous odour, and a somewhat acrid and bitter taste.

Impurities.—(1) Hemidesmus, which is cracked, but not annulated; (2) Almond meal (in powdered ipecacuanha), detected by odour of prussic acid when moistened.

Composition.—Ipecacuanha contains from $\frac{1}{4}$ to 1 per cent. of *emetin*, which is its active principle; *ipecacuanhic or cephaëlic acid*; starch, gum, etc. *Emetin*, $C_{20}H_{30}NO_5$, is a crystalline alkaloid, white, becoming yellow, odourless, bitter, comparatively insoluble in water, forming salts with acids which are readily dissolved in ordinary media.

Dose.—As an expectorant, $\frac{1}{2}$ to 2 gr.; as an emetic, 15 to 30 gr.

Preparations.

1. **Pulvis Ipecacuanhæ Compositus.** "Dover's Powder."—
Ipecacuanha, 1; Opium, 1; Sulphate of Potash, 8. A light fawn-coloured powder. *Dose*, 5 to 10 gr.

From Dover's Powder is prepared :

Pilula Ipecacuanhæ cum Scillâ.—Compound Powder of Ipecacuanha, 3; Squill, 1; Ammoniacum, 1; Treacle, q.s. *Dose*, 5 to 10 gr.

2. **Pilula Conii Composita.**—1 in 6. See *Conii Folia.*
3. **Trochisci Ipecacuanhæ.**—$\frac{1}{4}$ gr. in each. *Dose*, 1 to 3.
4. **Trochisci Morphiæ et Ipecacuanhæ.**—Ipecacuanha, $\frac{1}{12}$ gr.; Hydrochlorate of Morphia, $\frac{1}{36}$ gr. in each. See *Opium*, page 187.

5. **Vinum Ipecacuanhæ.**—1 in 20 of Sherry. *Dose*, 5 to 40 min.
as an expectorant; as an emetic, 3 to 6 dr.

ACTION AND USES.

1. IMMEDIATE LOCAL ACTION AND USES.

Externally.—Ipecacuanha powder is irritant, and even
pustulant, but is never used to produce this effect. Exposed
mucous membranes are similarly affected by it. If taken as
snuff it causes irritation of the nerves, sneezing, and reflex
mucous secretion; and the same effect follows its application in
smoke or spray to the pharynx, larynx, or lower air-passages;
in some persons it excites asthma. In the form of a spray of
the diluted vinum, or inhaled as the smoke of the burning
powder, it is used to relieve cough due to dryness or deficient
secretion of the throat and air passages.

Internally.—Reaching the stomach, ipecacuanha in very
small doses (gr. $\frac{1}{4}$) is a **gastric stimulant** doubtless increasing
the local circulation and secretion. It is therefore a useful
addition to bitter stomachic and tonic mixtures, and will even
arrest vomiting due to certain obscure conditions of the gastric
nerves. The compound powder is of the greatest value in
ulceration of the stomach, and some forms of dyspeptic vomiting.
In larger doses (15 to 30 gr.) it acts as an **emetic**, partly by a
direct effect upon the stomach, and partly by exciting the
vomiting centre in the medulla (indirect emesis). This important subject will be discussed under the heading of the
specific action.

In the intestines, ipecacuanha is still a stimulant, increasing
the flow of mucus; and in large doses an irritant. A remarkable tolerance of the drug is, however, readily established in
many persons suffering from **dysentery**, in which disease
ipecacuanha has the power of arresting the inflammatory action
in the bowel, checking the liquid and bloody evacuations, and
often effects a complete cure. For this purpose enormous doses
(30 to 90 gr.) are given, or large doses frequently repeated
(20 gr. every two hours).

2. ACTION IN THE BLOOD.

Emetin passes through the blood, from the alimentary canal
to the tissues, but is not positively known to affect it.

3. SPECIFIC ACTION AND USES.

Ipecacuanha (emetin) acts on the vomiting centre in the
medulla, *i.e.* is an **indirect emetic**, this effect being added to

the direct (gastric) action already mentioned. The ordinary doses (15 to 30 gr. of the powdered root, 3 to 6 fl.dr. of the vinum for adults) produce free evacuation of the stomach, and respiratory passages in 20 to 30 minutes, the dose often having to be repeated in 15 minutes, and the vomiting act probably occurring but once. But little nausea precedes, and moderate depression follows, the emesis. The circulation and respiration are disturbed and finally depressed by ipecacuanha, chiefly through the vomiting.

This drug is suitable as an emetic in cases where the necessity for evacuation of the stomach is not very urgent, and the subject is likely to be benefited by moderate but injured by great depression. It must not be given, therefore, in poisoning by alkaloids, such as morphia, but to children and weakly subjects in cases where the after effects of the drug will be also useful. It thus occupies a position amongst emetics between sulphate of zinc or copper and tartar emetic. Ipecacuanha may be used to empty the stomach in the early stages of sthenic fevers (less commonly than before); in cramp, whooping-cough, and the bronchitis of children, to expel membranes or mucous products from the air passages; and in acute dyspepsia with biliousness and heat of skin.

The skin is stimulated to increased secretion by ipecacuanha, which is used as a **diaphoretic**, especially combined with opium (Dover's Powder), in common colds, sore throat, and mild rheumatic attacks.

4. REMOTE LOCAL ACTION AND USES.

Emetin is excreted by the various mucous membranes, including those of the bronchi, the stomach, and bowels, and by the liver. On the bronchi it produces the same remote as immediate local action, namely, stimulation of the nerves, reflex cough, increased secretion, and, in large doses, even in-flammation of the mucous membrane and lungs. Ipecacuanha is thus an expectorant, increasing at once the expulsive acts, and the amount, that is, the liquidity, of the sputa. It is the most generally used of all this class of measures, being given in acute and chronic bronchitis, in phthisis, and in most cases of cough when the phlegm is scanty and tough. Important advantages of ipecacuanha are, that, if taken in excess, it causes sickness, which is often beneficial in the bronchitis of children; and that as a diaphoretic and moderate depressant of the circulation, *i.e.* a **sedative expectorant**, it controls the fever at the same time.

Acting remotely on the liver, this drug is a **direct chola-gogue**, increasing the secretion of bile; and has long been a favourite constituent of some purgative pills and aperient draughts for chronic biliousness and gouty dyspepsia.

Catechu Pallidum—Pale Catechu.—An extract of the leaves and young shoots of Uncaria Gambir. Prepared at Singapore and in other places in the Eastern Archipelago.

Characters.—In cubes, about an inch in diameter, externally brown, internally ochrey-yellow or pale brick-red, breaking easily with a dull earthy fracture. Taste bitter, very astringent, and mucilaginous, succeeded by slight sweetness. Entirely soluble in boiling water. The decoction when cool is not rendered blue by iodine.

Composition.—Catechu chiefly contains a crystalline bitter substance, *catechin* or *catechuic acid*, $C_{13}H_{12}O_5$, probably itself inactive; and *catechu-tannic* acid, the active principle, isomeric with it, and into which it is rapidly converted by boiling or by the action of saliva, with the development of a red colour. Both catechuic and catechu-tannic acids give a green precipitate with persalts of iron.

Incompatibles.—The alkalies, metallic salts, and gelatine.

Dose.—10 to 30 gr.

Preparations.

1. **Infusum Catechu.**—1 in 27. *Dose,* 1 to 2 fl.oz.
2. **Pulvis Catechu Compositus.**—Catechu, 4; Kino, 2; Rhatany, 2; Cinnamon, 1; Nutmeg, 1. *Dose,* 20 to 40 gr.
3. **Tinctura Catechu.**—1 in 8. *Dose,* $\frac{1}{2}$ to 2 fl.dr.
4. **Trochisci Catechu.**—1 gr. in each. *Dose,* 1 to 6.

ACTION AND USES.

Catechu acts like tannic acid, and is used for the same purposes. It is a favourite **astringent** application to sore throat in the form of the lozenge, and the compound powder and tincture are very commonly prescribed for diarrhœa.

Caffein. (*Not Officinal.*)—An alkaloid obtained from Coffea arabica, the coffee plant; from Thea sinensis, the tea plant (nat. ord., Camelliaceæ); from Ilex paraguayensis, Maté, or Paraguay Tea (nat. ord., Aquifoliacæ); and from Guarana. (*See* page 211.)

Characters.—Caffein, $C_8H_{10}N_2O_4$, occurs in fine long silky white prisms, soluble in water, with a bitter taste. It is a feeble base. Tea contains 1 to 4 per cent. of caffein ; coffee, 0·2 to 0·8 : maté, 1·2 ; guarana, 5 per cent. It is closely allied to theobromin, $C_7H_8N_4O_2$, being, in fact, methyl-theobromin, $C_7H_9(CH_3)N_4O_2$, which can be made synthetically.

Incompatibles.—Tannic acid, iodide of potassium, and salts of mercury.

Dose.—1 to 5 gr., or more.

Preparation.

Citrate of Caffein, in white needles. *Dose,* 1 to 5 gr.

ACTION AND USES.

1. IMMEDIATE LOCAL ACTION AND USES.

Coffee stimulates most of the digestive glands, being sialagogue, **stomachic** and slightly **laxative.** So far it is dietetically wholesome.

2. ACTION IN THE BLOOD, AND SPECIFIC ACTION AND USES.

Caffein is absorbed into the circulation unchanged; and acts chiefly upon the central nervous system. The **cerebrum is first stimulated,** causing the clearness of intellect, the removal of languor, and the sleeplessness, familiar after a cup of strong coffee. Larger doses cause a species of narcotism ; but there are great differences in this and other respects according to the individual and other circumstances. In the lower animals the spinal centres are simultaneously affected to such a degree that tetanic convulsions may occur, not unlike those caused by strychnia ; but in man these effects on the lower centres are quite subsidiary. The sensory and motor peripheral nerves are not certainly affected. The muscle curve is altered in character, and **muscular contraction seems more easily executed.** The **heart is first accelerated**—another familiar effect of a full cup of coffee ; it is then slowed and weakened. The blood pressure first rises and then falls. Respiration is temporarily increased, then depressed ; and death occurs in this way. Metabolism is probably somewhat increased, and the temperature raised. In all these respects habit markedly reduces the influence of coffee.

Coffee and caffein may be used as a nervine stimulant and restorative in ordinary conditions of fatigue. Megrim is frequently relieved by either. Large doses must be avoided.

3. REMOTE LOCAL ACTION AND USES.

Caffein is excreted unchanged in the bile and urine, the latter presenting the characteristic odour of the substance.

In passing through the kidney, it appears to stimulate the cells, at least in some subjects, and acts as a **diuretic.** Citrate of caffein is thus a powerful, but somewhat uncertain, remedy in dropsy, whether cardiac or hepatic. It is best given after, and then along with, a stimulant diuretic, such as digitalis; for a short time; and in moderate doses.

VALERIANACEÆ.

Valerianæ Radix — VALERIAN ROOT. — The dried root of Valeriana officinalis. From plants indigenous to and also cultivated in Britain. Collected in autumn, wild plants being preferred.

Characters.—A short yellowish-white rhizome, with numerous fibrous roots about two or three inches long; of a bitter taste, and a penetrating odour, agreeable in the recent root, becoming fetid by keeping; yielding volatile oil and valerianic acid when distilled with water.

Substances resembling Valerian : Serpentary, Arnica, Veratrum Viride, known by odour.

Composition.—The active principles of valerian are a *volatile oil* and *valerianic acid.* The oil consists of a terpen *valerene,* $C_{10}H_{16}$, and an oxygenated oil, *baldrian camphor,* $C_{12}H_{20}O$. Valerianic acid, $C_5H_{10}O_2$, occurs in a large number of other plants, and in cod-liver oil; and can be derived from fousel oil (amylic alcohol, or valeryl-aldehyd), $C_5H_{12}O$, by oxydation. It is a colourless oily fluid, with a powerful odour and acid burning taste; soluble in 30 parts of water, and freely in alcohol and ether.

Dose, in powder.—10 to 30 gr.

Preparations.

A. *Of Valerian Root :*

1. **Infusum Valerianæ.**—1 in 36. *Dose,* 1 to 2 fl.oz.
2. **Tinctura Valerianæ.**—1 in 8 of Proof Spirit. *Dose,* 1 to 2 fl.dr.
3. **Tinctura Valerianæ Ammoniata.**—1 in 8 of Aromatic Spirit of Ammonia. *Dose,* ½ to 1 fl.dr.

B. *Containing Valerianic Acid :*

Sodæ Valerianas.—Valerianate of Soda. $NaC_5H_9O_2$.

Source.—Made by (1) distilling Amylic Alcohol with Sulphuric Acid and Bichromate of Potash; (2) saturating the distillate with Liquor Sodæ, evaporating, liquefying, and

Made in the USA
Monee, IL
07 July 2026

56551326R10166